Neurophysiology: The Fundamentals

NEUROPHYSIOLOGY:
The Fundamentals

Robert A. Lavine, Ph.D.

Associate Professor of Physiology and of Neurology
The George Washington University
School of Medicine and Health Sciences
Washington, D.C.

The Collamore Press
D.C. Heath and Company
Lexington, Massachusetts
Toronto

Published simultaneously in Canada and the United Kingdom

Printed in the United States of America

International Standard Book Number: 0–669–04343–5

Library of Congress Catalog Card Number: 80–2611

Library of Congress Cataloging in Publication Data

Lavine, Robert A.
  Neurophysiology: The fundamentals.

  Bibliography: p.
  Includes index.
  1. Neurophysiology.   I. Title.   [DNLM:
1. Nervous system—Physiology.   WL 102 L412n]
QP361.L28   612'.8   80–2611
ISBN 0–669–04343–5   AACR2

Contents

Preface and Acknowledgments   *vii*

1. **Structure and Function**   *How the nervous system is designed to communicate information and control activity*   1

2. **The Nerve Impulse**   *How nerve cells become excited and conduct impulses*   17

3. **Sensory Reception**   *How receptors convert energy from the environment into nerve impulses*   31

4. **Synaptic Transmission**   *How signals pass from neuron to neuron and from neuron to muscle fiber*   37

5. **Somatic Sensation**   *How the nervous system converts touch, pressure, joint position, temperature, and painful stimuli into sensations*   55

6. **Auditory and Vestibular Function**   *How neural mechanisms function in hearing and balance*   69

7. **Vision**   *How the eye works and how the visual system converts light into sensed visual images*   87

8. **Chemical Senses**   *How the nervous system responds to chemicals to produce sensations of taste and smell*   107

9. **Posture and Movement**   *How motor systems control the skeletal muscles*   113

10. **Autonomic Function**   *How sympathetic and parasympathetic activity help control the internal environment*   133

11. Cerebral Physiology and Higher Functions   *How cerebral activity relates to wakefulness and sleep, language, memory, and other complex processes*   139

Glossary   157
References   166
Suggested Readings   167
Index   169

Preface and Acknowledgments

The student of medicine, biomedical science, or allied health is asked to master an enormous amount of material. Presented as textbooks, lectures, handouts, and examinations, the material may appear bewildering in variety and overwhelming in quantity. In an attempt to deal with it, the student may passively remember bits and pieces but never master the essential concepts of the subject matter. This is especially true in neurophysiology, which has a terminology of its own and combines elements of anatomy, physics, chemistry, psychology, neurobiology, and neurology, as well as physiology. Textbook treatments are often wordy and replete with poorly defined abstractions.

This book is designed to assist the beginning student in several ways. Each chapter contains a concise, concrete, and well-illustrated discussion of a particular topic in neurophysiology. The aim is to be clear about essentials rather than encyclopedic in scope. The chapters can stand by themselves as an introduction to neurophysiology, especially for those in allied health courses and readers who do not need more detail, or can be followed by more detailed sources. Examples from everyday life, medicine, and laboratory research contribute to the meaningfulness of the material. Each chapter ends with a set of review exercises—items to describe, explain, list, draw, and so on. The best way to learn is to do something with the subject matter, whether by experiments, clinical practice, or, as with the exercises, by describing the material to yourself and others. The exercises can serve as a guide to study in which you participate actively, as well as a basis for review before examinations. If you can satisfy the objectives of these exercises, you will have a mastery of a basic core of neurophysiology.

It is a pleasure to acknowledge a number of people who contributed to the production of this book. Barbara Lavine provided encouragement and moral support, as well as practical assistance with several figures and the index. Richard A. Kenney, Ph.D., encouraged me to begin and carry through the enterprise and provided illuminating discussions of the material. Reading and helpful criticisms of all or part of the manuscript were provided by Dr. Kenney and by Rebecca Anderson, Ph.D., Marie M. Cassidy, Ph.D., Paul Mazel, Ph.D., Lawrence Rothblat, Ph.D., Paul F. Teychenne, M.D., and Penelope S. Myers, M.A. These colleagues were aware of the limitations imposed on each topic by the introductory nature of the book, and the author takes responsibility for the accuracy of the outcome. Sharon Reutter assisted with library research and editing. Medical, graduate, and allied health students provided a responsive audience for which this material was developed.

Special thanks are due to Virginia L. Schoonover and Judith Guenther for their expert rendering of most of the figures in the book; to Debra Fink, Selma Klein, and Lisa Landau for their skilled typing of the manuscript; and to Sarah Boardman and her editorial colleagues at The Collamore Press for their advice and support in carrying the book to completion.

Neurophysiology: The Fundamentals

1. Structure and Function

*How the nervous system is designed to communicate
information and control activity*

The Cellular Structure of the Nervous System

Neurons

The building blocks of the nervous system are the nerve cells or neurons. Neurons are specialized to generate and carry information by means of electrical and chemical signals. A typical neuron, as shown in figure 1–1, consists of an enlarged area called the *cell body* (or soma), which contains the nucleus and much of the metabolic machinery of the cell; the *dendrites*, several short, thin extensions that divide into branches like a tree and receive information from other neurons; and the *axon*, a specialized thin extension that transmits neural signals for a relatively long distance, after which it divides into several branches that end on other cells. The axon shown in figure 1–1 is covered by segments of an insulating lipid material, the *myelin sheath*, and arises from the cell body at the *axon hillock*, a cone-shaped region, followed by the *initial segment*, a short length of bare axon before the myelin-covered length of axon begins.

Variations on this basic scheme can be found. For example, the axon may lack a myelin sheath; the axon may be short; or the cell body (in spinal sensory neurons) may be placed to one side of the axon.

Synapses

The axon of one neuron (or its branch) communicates with another neuron across a microscopic gap at a junction called a *synapse*. The axon terminal is typically enlarged and contains chemical compounds in numerous membrane packages called *vesicles*. When the axon terminal forms a synapse upon the cell body (soma), dendrite, or axon of another neuron, the synapses are termed *axosomatic*, *axodendritic*, or *axoaxonal*, respectively. The axon may also terminate on a muscle cell in a synapse called a *neuromuscular junction*, or on a gland cell.

Receptors

Sensory receptors are structures that convey information about the outside world and the interior of the body to the spinal cord and brain (figure 1–2). Anatomically, they include a wide variety of types: free nerve endings; nerve endings associated with capsules, hair follicles, muscle fibers, or other cells; and highly modified nerve cells, such as the rods and cones of the retina. Each sensory receptor is specialized to receive a particular type of stimulus (touch, light, sound, chemicals, or others) and translate it into an electrical signal. After modification, these signals are carried into the spinal cord and brain by sensory nerve axons associated with the receptors. Thus, receptor activity is the initial input stage of the nervous system.

Effectors

Moving, talking, and facial expression are controlled by *muscle fibers*, one form of effector

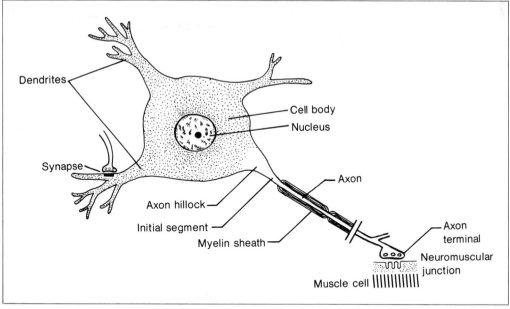

A

Figure 1–1. (A) *Diagram of a typical neuron, showing cell body, dendrites, synapse, axon with myelin sheath and axon terminal, and a neuromuscular junction.* (B) *Photomicrograph of an alpha motoneuron in ventral horn of cat spinal cord. Four dendrites emanate from the cell body. Arrow points to the axon, at the junction of the initial segment with the myelinated portion.* (C) *Photomicrograph of a bundle (or fascicle) of myelinated axons, forming a small peripheral nerve, shown in cross-section. Dark borders around each axon are myelin sheath. White circle in center of bundle is a blood vessel. The nerve shown is cranial nerve IV (trochlear) of the rat, which supplies eye muscle fibers; it contains an average of 270 axons, the largest of which are 10 μ in diameter. Courtesy of Dr. James M. Kerns.*

(figure 1–2). In addition, the internal environment of the body, including blood pressure, metabolic rate, and body temperature, is controlled not only by muscle fibers but also by the secretions of *glands*, another form of effector. The effectors are the means by which the nervous system exerts an effect on both the outside and inside environments. Their activity represents the final output of the nervous system.

General Plan of the Nervous System

The two main divisions of the nervous system are the *peripheral nervous system* (*PNS*) and the *central nervous system* (*CNS*) (figure 1–3).

The Peripheral Nervous System

The peripheral nervous system (PNS) consists of the *peripheral nerves*, bundles of axons (nerve fibers) that function to carry signals back and forth between peripheral organs (receptors and effectors) and the central nervous system. These nerve fibers are subdivided according to the peripheral organs on which they end: *somatic* nerve fibers supply skin, skeletal muscles, tendons, and joints, and visceral nerve fibers supply the gut and other *visceral* organs. Classified according to function, *afferent* nerve fibers carry sensory information to the central nervous system (input), while *efferent* nerve fibers carry motor-command signals from the central nervous system to the effectors (out-

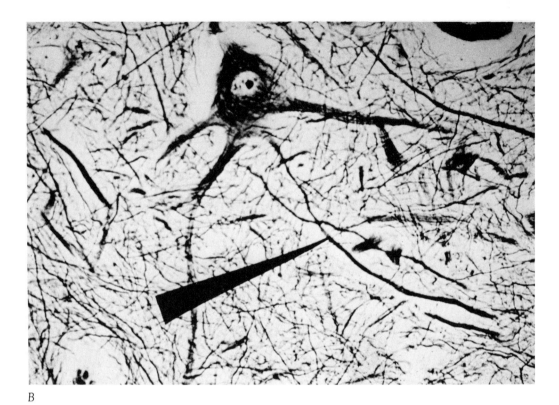

B

C

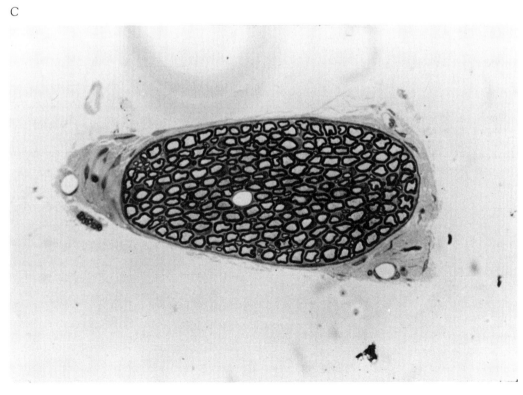

3

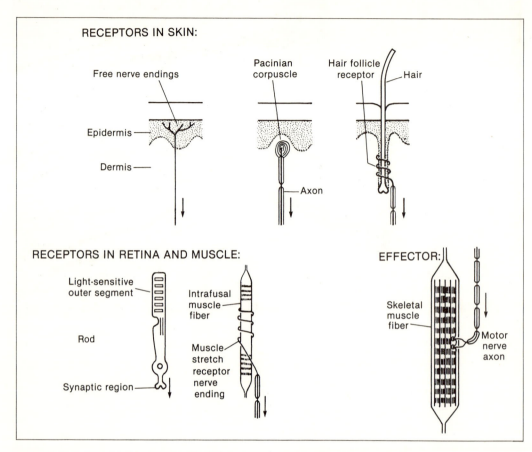

Figure 1–2. Several types of sensory receptors and an effector. The three receptors on top are associated with skin; the rod is in the retina; and the muscle stretch receptor is associated with specialized (intrafusal) muscle fibers in skeletal muscle. A larger (extrafusal) muscle fiber that acts as an effector is shown at lower right. Arrows indicate direction of neural signals.

put). There are then four types of peripheral spinal nerve fiber:

1. Somatic afferent, carrying sensory information from skin, skeletal muscles, tendons, and joints
2. Somatic efferent, carrying motor-command signals to skeletal muscles
3. Visceral afferent, carrying sensory information from the visceral organs
4. Visceral efferent, carrying motor-command signals to visceral effectors (smooth muscle, cardiac muscle, and glands); these nerves are also called the *visceral motor* or *autonomic nervous system*, and are further subdivided into sympathetic and parasympathetic systems.

Figure 1–3. (A) Diagram of nervous system as seen from dorsal surface (not to scale), divided into peripheral nervous system (PNS) and central nervous system (CNS), consisting of spinal cord and brain. Spinal segments are C = cervical, T = thoracic, L = lumbar, S = sacral, and Co = coccygeal. Only somatic afferent and efferent fibers are shown. MN, motoneuron (alpha motoneuron controlling muscle fibers in arm). Ascending fibers shown carry sensory information from spinal cord to somatic sen-

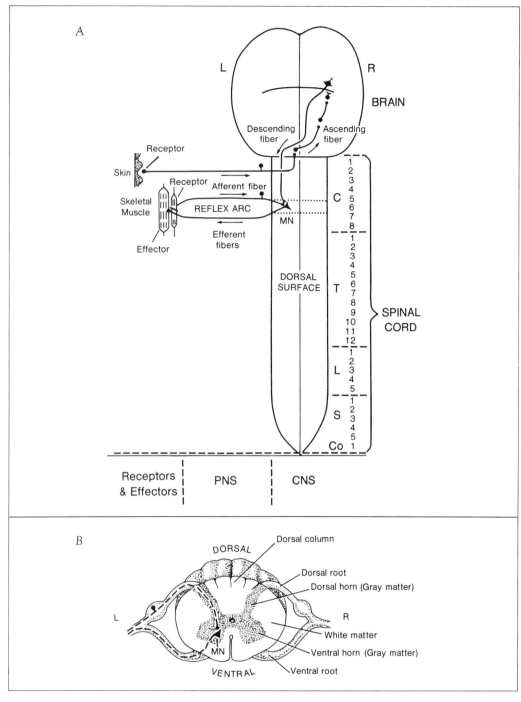

A

L   R

BRAIN

Descending   Ascending
fiber   fiber

Receptor

Skin

Receptor   Afferent fiber

Skeletal
Muscle

REFLEX ARC

Effector

Efferent
fibers

MN

DORSAL
SURFACE

C
1
2
3
4
5
6
7
8

T
1
2
3
4
5
6
7
8
9
10
11
12

L
1
2
3
4
5

S
1
2
3
4
5

Co 1

SPINAL
CORD

Receptors
& Effectors

PNS

CNS

B

DORSAL

Dorsal column

Dorsal root

Dorsal horn (Gray matter)

L   R

White matter

Ventral horn (Gray matter)

MN

VENTRAL

Ventral root

sory cortex of brain; descending fiber shown carries
motor information from motor cortex of brain to
spinal motoneuron. (B) A segment of spinal cord,
corresponding to section within dotted lines in A.
White matter contains axons, gray matter contains
cell bodies, dendrites, and synapses. Within gray
matter, dorsal and ventral horns are shown; within

white matter, dorsal column is shown. Dorsal root
contains afferent fibers, and ventral root contains efferent fibers originating in motoneurons; dashed lines
indicate individual fibers corresponding to those
shown in the reflex arc in A. Enlarged area (ganglion) of dorsal root contains cell bodies of afferent fibers.
Dorsal and ventral roots join to form peripheral nerves.

Table 1–1. Cranial Nerves and Their Major Functions

Nerve	Afferent Fiber Functions	Efferent Fiber Functions
I Olfactory	Smell	
II Optic	Vision	
III Oculomotor		Eye movement Constriction of pupil Accommodation of lens
IV Trochlear		Eye movement
V Trigeminal	Somatic sensation from part of face and head	Middle-ear reflex (via tensor tympani muscle) Chewing (mastication)
VI Abducens		Eye movement
VII Facial	Taste from anterior two-thirds of tongue	Middle-ear reflex (via stapedius muscle) Facial expression Tearing (lacrimation) Salivation
VIII Auditory (vestibular)	Hearing Maintenance of equilibrium	
IX Glossopharyngeal	Taste from posterior one•third of tongue Blood pressure information from cervical baroreceptors Chemical information from cervical chemoreceptors	Salivation Swallowing
X Vagus	Taste from part of tongue and pharynx Visceral sensation from pharynx, thorax, and abdomen Blood pressure information from thoracic baroreceptors Chemical information from thoracic chemoreceptors	Salivation Swallowing Slowing of heart rate Secretion and patency of bronchioles Production of surfactant in lungs Secretions and movements of abdominal digestive organs
XI Accessory		Swallowing Turning head Elevating shoulder
XII Hypoglossal		Tongue movement

The peripheral nerve fibers enter and exit the spinal cord via a series of *roots:* the dorsal roots carry afferent nerve fibers into the cord and the ventral roots carry efferent nerve fibers from the cord. The roots are labeled according to associated segments of the vertebral column: cervical (C), thoracic (T), lumbar (L), sacral (S), and coccygeal (Co). They are numbered within each segment (C1 through 8; T1 through 12, L1 through 5, S1 through 5, and Co-1). The arm is innervated by spinal nerves C4 through T1 and the leg by nerves L2 through S3. The peripheral nerves directly connected to the brain are cranial nerves of which there are twelve pairs. Their names and major functions are listed in table 1–1.

The Central Nervous System

The central nervous system (CNS) consists of the brain and spinal cord (figure 1–3A), each of which is enclosed in a bony covering, the brain within the skull and the spinal cord within the vertebral column. Both are divided at the midline into a left and right half and have surfaces labeled *dorsal* (toward the back) and *ventral* (toward the front).

Within the CNS are *white* and *gray areas*. The white areas contain bundles of nerve axons (also called *nerve fibers*); their color is imparted by the myelin sheath around the nerve fibers. There are several anatomical terms for specific bundles of nerve fibers within the CNS, such as *tract, column, fasciculus, lemniscus,* and *radiation*. The gray areas contain the cell bodies of neurons, their dendrites, and associated synapses.

In cross-section, the gray areas in the spinal cord form a butterfly-shaped area in the center of the cord, surrounded by a ring of white matter (figure 1–3B). The white matter contains bundles of axons passing up and down through the cord connecting areas of the cord with each other and with the brain, while the central gray area contains numerous synaptic regions. Thus, a sensory signal may enter the cord through a dorsal-root nerve axon that (1) travels up to the brain in an ascending column or tract or (2) enters the gray area to synapse on other nerve cells. Within the brain, gray areas consist of cell bodies and synapses organized into *nuclei* (clusters) and *layers* (in the surface of the cerebrum and cerebellum).

The brain is the highest level of the CNS. It is the organ ultimately responsible for movement, perception, and thought. The brain has four major divisions (figure 1–4):

1. The *brainstem* partially resembles an extension of the spinal cord upward into the head. It is composed of the medulla, pons, and midbrain. Like the spinal cord, the brainstem contains white bundles of nerve fibers (called lemnisci, tracts, and so on) carrying signals both up and down, gray nuclei containing cell bodies and synapses, and the roots of peripheral nerves—the cranial nerves—arranged in a more complex fashion than the spinal nerves and carrying

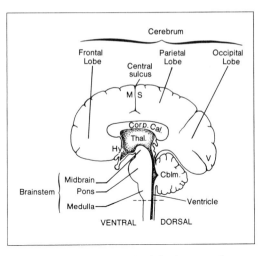

Figure 1–4. *The right half of the brain in diagrammatic form, as seen from the left (a midsaggital section). Abbreviations:* **Cblm.** = *cerebellum,* **Thal.** = *thalamus,* **Hy.** = *hypothalamus,* **Corp. Cal.** = *corpus callosum,* **M** = *motor cortex,* **S.** = *somatic sensory cortex,* **V.** = *visual cortex. Temporal lobe not shown. The ventricles contain cerebrospinal fluid. Dashed line shows juncture of spinal cord and brain.*

information from the eyes and ears as well as other sources. Within the core of the brainstem a network of nerve cells called the *reticular formation* is essential in regulating wakefulness and sleep, respiration, and cardiovascular function.

2. The *cerebellum* is a globular structure connected to the dorsal surface of the brainstem by bundles of nerve fibers. Its surface is covered with folds, so that it partly resembles a ball of twine. A major function of this organ is to correct errors so that movement proceeds in a smooth and controlled manner—so that the hand, for example, can reach for and pick up a cup without spilling the contents.

3. The *thalamus* and *hypothalamus*, together called the *diencephalon*, are gray areas containing nuclei (clusters of cells) located above the brainstem and near the center of the brain, surrounding the walls of the third ventricle. The thalamus is a major relay

station for information traveling up to the cerebral cortex from other areas. The hypothalamus (ventral to the thalamus) is a control center for visceral functions, including, for example, changes in cardiovascular function during anger and changes in thyroid gland secretion during cold exposure.

4. The *cerebrum,* divided at the midline into two *cerebral hemispheres* connected by a thick band of axons called the *corpus callosum,* is the largest and most visible part of the human brain. Comparison of the human cerebrum with that of lower vertebrates suggests a great increase in size and function in the course of evolution. The outer surface or cerebral *cortex* is known as the gray matter because of its layers of nerve cells. The surface area of the cortex is increased by its numerous folds (gyri), separated by grooves (sulci and fissures). Beneath the cortex is the white matter, consisting of nerve fibers traveling to and from the cells in various parts of the cortex. Each cerebral hemisphere is divided into four lobes (figure 1–5): the *occipital* (at the back of the head), the *parietal* (top), the *temporal* (sides), and the *frontal* (at the front of the head, behind the forehead). The cortex contains motor areas that govern voluntary movement and keep reflex movements under control, sensory areas that register incoming information from sensory receptors (with separate areas for vision, hearing, body sensation, and so on), and association areas that associate information from several sources and carry out such higher functions as the understanding of speech and the recognition of objects. Within the white matter are gray areas that contain clusters of cells and synapses and are divided into nuclei. Between the sides of the thalamus and the cerebral cortex are the *basal ganglia,* containing several nuclear structures that help to regulate posture and movement. Below the surface of the tip of the temporal lobe is the *amygdala,* containing nuclei that can influence such behaviors as fighting and eating; the amygdala is part of the limbic system, a group of primitive structures within the cerebrum.

Neural Signals

The nervous system uses signals to transmit information. Information is transmitted from the outside world into the nervous system (by means of sensory processes), from one part of the nervous system to another, and from the nervous system back to the outside world (by means of motor processes). The signals are carried by the nerve cells, including their extensions, the dendrites and axons. In everyday life, signals include traffic lights, Morse code, and the electronic pulses within computers. In the nervous system, signals can be divided into action potentials, graded receptor potentials, graded synaptic potentials, and chemical transmitter release.

Action Potentials

Like one of the voltaic cells in a battery, the nerve cell at rest is electrically charged across its membrane. The charge gives rise to an electrical potential difference, a *membrane potential,* measured in *millivolts* (thousandths of a volt, abbreviated mV). The magnitude of this potential difference ranges from 55 mV to 90 mV, with the inside of the cell negative compared with the outside. This membrane potential is altered by the *action potential,* a brief reversal of polarity that sweeps along the nerve axon (figure 1–6).

The action potential is triggered by electrical or other stimuli above a certain strength. When neural signals need to be transmitted over long distances, such as from one end of an axon to another, the signals used are action potentials.

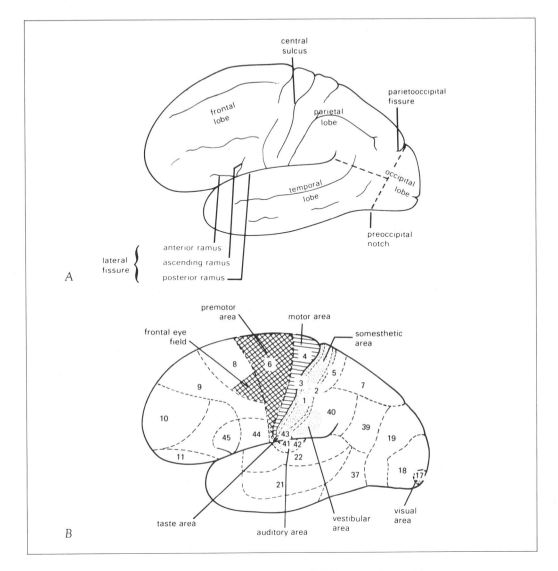

Figure 1–5. Lateral views of the left cerebral hemisphere. (A) Division into lobes; sulcus and fissure are terms for grooves in the cortex; the lateral fissure has three rami or branches. (B) Subdivision into areas based on function (auditory, visual, and so on) and on the structure of cell layers, or cytoarchitecture, with numbers given by Brodmann. From Barr, M.L. The Human Nervous System: An Anatomic Viewpoint (3rd ed.). Hagerstown, Md.: Harper & Row Publishers, Inc., 1979. Reprinted with permission.

Graded Receptor Potentials

In neurons specialized to receive sensory stimuli (receptors), stimuli can produce a change in membrane potential called a *receptor potential*. The change is graded in size, depending on the stimulus. For example, in receptors sensitive to pressure on the skin, the greater the pressure the greater the potential change (figure 1–6).

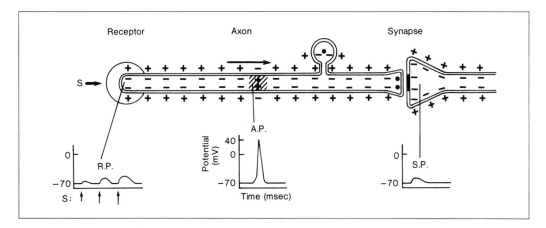

Figure 1–6. Neural signals. **S** = *stimulus (arrows below graph indicate stimuli of three intensities),* **R.P.** = *receptor potential,* **A.P.** = *action potential,* **S.P.** = *synaptic potential.*

Graded Synaptic Potentials

Synapses are specialized regions where two neurons communicate across a microscopic gap. At synapses, neurons respond to signals arriving from other nerve cells by generating *synaptic potentials.* Like receptor potentials, these are graded in size according to the strength of the stimulus. Another type of synapse is the junction between nerve and muscle cells.

   Graded potentials, whether in receptors or synapses, carry information only over very short distances, such as within a nerve cell body.

Chemical Transmitter Release

The synaptic gap between neurons is generally crossed by chemical transmitters (for example, acetylcholine or norepinephrine). When an action potential reaches the end of a nerve axon, it causes the release of a chemical transmitter stored in the axon ending, which diffuses across the synaptic gap to the next nerve cell. Chemical transmitters, therefore, carry information from cell to cell.

The Timing of Neural Activity

Since neurophysiology concerns sequences of activity in which information is transmitted by precise electrical and chemical signals, time is an important dimension. Time is generally measured in milliseconds (thousandths of a second, abbreviated msec). There are four basic parameters that can be measured: duration, velocity, intervals, and frequency.

Duration

The duration of an action potential is easy to remember—about 1 msec. During this brief time, the membrane state changes from resting to active and then returns to resting. Graded potentials generally last longer; synaptic potentials in the spinal cord, for example, have durations of approximately 12 msec.

Velocity

The action potential is conducted along axons at velocities ranging from about 1 to 100 m/sec, depending on the structure and function of the axon. For example, among different fibers in peripheral nerves, those conducting at a velocity of about 100 m/sec may carry muscle-length information; those conducting at about 50 m/sec may carry touch and pressure information; and those conducting at about 1 m/sec

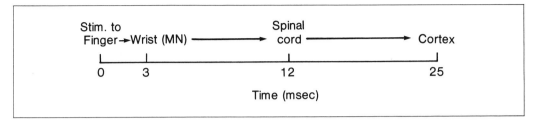

Figure 1–7. The time sequence of neural responses along a sensory pathway. **MN** = *median nerve. (Neural responses at cervical spinal cord and somatic sensory cortex are averaged evoked potentials, obtained by computer processing of electrical signals recorded with disc electrodes on the skin).*

may carry pain and temperature information. These velocities are slowed down in certain diseases and therefore can be useful in clinical diagnosis.

Intervals

Time intervals in neural transmission can be precisely measured, even from outside the body. For example, recording electrodes on the skin can be used to measure the intervals between an electrical pulse applied to the fingertip and the action potentials at several stages of the sensory pathway. The action potential is first conducted from the digital nerves in the finger to the median nerve, from which it can be recorded at the wrist after an interval of about 3 msec (figure 1–7). After about 25 msec, a neural response can be recorded over the appropriate area of the cerebral cortex. When the time interval and the distance between two points along the peripheral nerve are determined, distance can be divided by time to calculate the conduction velocity. For the pathway to the cerebral cortex, the time interval includes synaptic delays as well as nerve-conduction time.

In a single nerve cell, the time intervals between action potentials depend on two factors: the properties of the nerve cell and the strength of the stimulus affecting it. A stronger stimulus generally causes a shorter interval between action potentials.

Frequency

Frequency can be defined as the number of events per second. In the case of action potentials in a single nerve cell, the frequency is the number of action potentials per second and is the inverse of the mean interval between action potentials; the frequency is high when the intervals are short. Thus, a stronger stimulus generally causes a higher frequency of action potentials. Another way of approaching this is to say that the frequency of action potentials generally acts as a code, transmitting information about the strength of a stimulus. (Inhibition and other factors modify this code in ways that will be discussed in later chapters.)

In the case of electrical waves recorded from the brain as part of the electroencephalogram (EEG), the frequency is the number of waves or cycles per second (c/s). A well-known type of EEG pattern consists of alpha waves, occurring at about 10 c/s. Different levels of sleep or wakefulness are associated with higher or lower EEG frequencies.

The components described thus far can be further organized according to three overall processes: sensory, motor, and higher.

Sensory Processes

A primitive animal creeping along the ocean floor has a reasonable chance of survival if it can escape when brushed by a larger predator, and a greater chance if it can sense the shadows and vibrations set up by the predator at a distance. Similar functions are now carried out by the human sensory processes—body sensation, vision, hearing, and others. Taking somatic sensation (body sensation, including touch and pressure) as an example, we can look at the sequence of activities that allow us to sense the outside world so that we can deal with it.

Transduction

Pressure on the skin (after someone hands you a book, for example) is a common type of sensory stimulus. Specialized receptors in the skin respond to the pressure with a graded receptor potential (an electrical signal). The transformation of one form of energy (mechanical) into another (electrical) is called *transduction*. Transduction is the fundamental step that permits the nervous system to react to events in the outside world.

Conduction

Action potentials are generated at the peripheral (receptor) end of the sensory axon and are then conducted at a characteristic velocity to the spinal cord or brain. Because of unpredictable fluctuations of neuron activity, some information in each axon is lost. This is compensated for by a number of parallel axons transmitting the same message so that information lost in one axon is transmitted in the others.

Integration

Chemical transmitters, released from several axons, affect the next neuron in the sensory system (figure 1–8). This allows the neuron to integrate (put together) signals from several sources. Such integration is a function of the synapse.

Conduction along axons and integration at synapses occur at successive levels of the sensory pathway until the cerebral cortex is

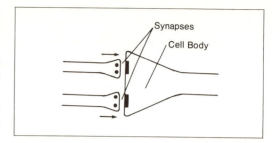

Figure 1–8. Information arriving at synapses from two axons is combined by cell body, an example of synaptic integration.

reached. At higher levels, some neurons respond as if they detect certain complex features of the stimulus, such as, the angle of a line or the change in pitch of a tone. This feature detection is one important result of the ability of synapses to integrate several incoming signals.

Specificity

Some receptors in the somatic sensory system are specialized to process pressure; receptors in the ear, sound waves; and the rod receptors in the retina, light. In all three cases, the outcome is a graded receptor potential. Each receptor is most sensitive to a specific type of external energy stimulus (figure 1–9). Each receptor also is connected to a specific sensory system of nerve cells in the brain and spinal cord and transmits neural signals through that system. The different sensations produced (pressure, sound, light) are due to such specific connections.

Motor Processes

Skeletal muscle contractions either hold the body in a relatively fixed posture (sitting, standing, waiting for a tennis ball) or cause it to move (writing, lifting, swinging at the tennis ball). Both posture and movement are motor processes, controlled by somatic efferent nerve

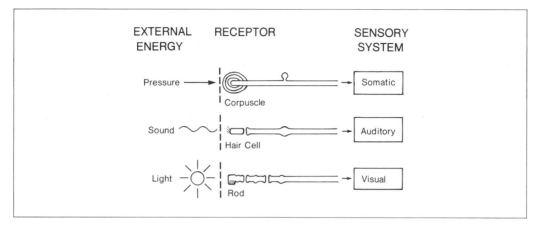

Figure 1–9. Specificity of receptors.

fibers and by the motor systems in the CNS that govern these nerve fibers.

Consider some of the motor processes involved in picking up a book. A convenient structure on which to focus attention is one of the spinal motoneurons supplying the arm muscles.

Integration

The motoneuron receives information from numerous sources over thousands of synapses. The sources include the muscle stretch receptors, which provide information about the length of the muscle, and descending fibers from various motor centers in the brain (figure 1–3).

The pathway from the muscle-stretch receptor to the spinal motoneuron to the muscle effector is an example of a *spinal reflex*, which is an automatic motor response utilizing a circuit built into the spinal cord. The effect of this reflex would be to support the weight of the book, opposing the force of gravity. The pathway from the brain motor centers to the spinal motoneuron is an example of *supraspinal motor control*. It governs the strength of the spinal reflex and any voluntary, purposeful movement, such as raising the book to eye level.

The information arriving over these and other channels is integrated by the spinal motoneuron (figure 1–8) before it produces one or more action potentials as an output signal to the muscle effector.

Conduction

The action potentials are then conducted along the motoneuron axon (a somatic efferent fiber) to the arm muscle it supplies. Each axon supplies a number of separate muscle fibers within the muscle. At the nerve ending a chemical transmitter is released, crosses the neuromuscular junction, and alters the electrical potential of the muscle membrane.

Transduction

The electrical potential change within muscle then leads to muscle contraction (a mechanical change). This change from one form of energy to another is a form of transduction, which is the reverse of what occurs in receptors that transduce mechanical into electrical energy.

Autonomic Nervous System

Posture and movement are governed by contractions of the skeletal muscles. On the other hand, the internal state of the body, including heart rate, blood pressure, pupil size, and the secretions of the digestive tract, is regulated largely by activity in smooth muscles and exocrine glands. These effectors are governed by the visceral efferent fibers in the autonomic nervous system.

The autonomic system is divided into a *sympathetic* division, important in protecting against danger and cold, and a *parasympathetic* division, partly involved in digestion and other relaxing activities. Since both systems are important for regulation of the internal environment of the body, an imbalance, such as excessive sympathetic activation due to a tumor in the adrenal medulla gland, can seriously impair this regulation. Visceral efferent activity is commonly considered involuntary, but under certain conditions people can learn to affect variables such as their heart rate.

Higher Processes and Cerebral Activity

A number of psychological processes are not purely sensory or motor and are often called higher functions or higher processes because of their complexity and the belief that they are better developed in humans than in lower animals. These processes include consciousness (ranging from deep sleep to alertness), language, and memory. Their partial association with cerebral activity can be shown by electroencephalography and other physiological techniques as well as by their impairment in disorders affecting the cerebrum.

Consciousness and the Electroencephalogram

The electroencephalogram (EEG) can be recorded with metal disc electrodes placed on the scalp and connected to electronic devices that amplify and display the minute electrical waves emanating from the cerebral cortex. The frequency of these waves is related roughly to levels of consciousness or alertness, from about 2 c/s during a deep sleep up to about 20 c/s during excitement. Clinically, EEG changes can be found during coma and epilepsy, as well as other conditions. In addition, computer processing of the EEG can provide recordings that reflect the brain's electrical response to a sensory stimulus, such as a click or a visual pattern. These responses, called *averaged evoked potentials,* can be altered by physiological, psychological, and pathological processes.

Language and Left-Right Cerebral Differences

People communicate with each other by means of language, both spoken and written. The neural basis of language has been demonstrated mostly by the study of patients with some sort of language disorder. Language problems can result from deafness (a sensory disorder) or paralysis of the tongue and throat (a motor disorder), but in addition, there are language disorders, called *aphasias,* that are due to disturbances in higher processes. In the majority of such patients, lesions have been found in the left cerebral hemisphere, leading to the conclusion that the left hemisphere is dominant for language in the majority of the population (over 90%). Particular areas within the parietal, temporal, and frontal lobes are related to different forms of aphasia; one form involves an inability to understand speech, another involves an inability to produce it. Studies of patients with severed connections between cerebral hemispheres (split-brain patients), methods of activating and recording from each hemisphere separately, subtle structural differences between left and right areas of cerebral cortex, and a clinical test in which left and right hemispheres can be anesthetized separately for a few seconds are extending our understanding of the cerebral control of language.

Memory

To students of the biomedical sciences who must remember enormous amounts of material, memory has more than theoretical interest. How much can it hold, and how long does it last? One experimental approach is to present a list of items (words or numbers) and ask someone to repeat them in the correct order immediately afterward. For example, you might read each of these lists of numbers, one digit at a time, and ask someone to repeat them immediately after each list:

List 1: 6 4 1 9
List 2: 1 8 2 5 6 4
List 3: 2 1 8 6 4 9 3 2 8 9 7

You will probably find that while most adults can repeat List 1, containing 4 digits, and List 2, containing 6 digits, they cannot do so for List 3, containing 11 digits. This is a test for *short-term memory* and demonstrates that its capacity is limited. *Long-term memory*, which lasts hours, days, or years, has no clear limits in its capacity and seems to be organized on different principles. Moving material to be memorized from short-term memory to long-term memory is a most important step. Interestingly, this step seems to be associated with the function of a particular area of the brain—the hippocampus, a cellular structure within the temporal lobe.

Memory is closely associated with learning in human behavior. Both can be demonstrated at a simple level. If you snap your finger once, twice, or three times, a child may be startled. After 20 times you will probably be ignored. The change in motor response, from startle to boredom, is evidence that the repeated sounds have been stored in memory, and the listener has learned that they are harmless. In other words, there has been a modification of behavior due to experience. Learning and remembering lists of numbers, or even learning neurophysiology, are also modifications of be-

havior due to experience, but on a more complex and advanced level.

Review Exercises

1. Diagram a neuron, labeling the cell body, dendrites, axon, and synapses.
2. Describe the role of receptors and effectors.
3. List and define the major components of the peripheral nervous system and the central nervous system.
4. Compare the white and gray areas of the CNS.
5. List in outline form the major structures within the brain.
6. Compare action potentials with graded potentials. Compare their roles in neuronal communication.
7. List approximate values for: (1) the duration of an action potential, (2) the conduction velocity of a typical somatic afferent axon responding to touch, and (3) the time interval between a stimulus applied to the fingertips and a response in the cerebral cortex.
8. Define transduction in a sensory receptor.
9. Give an example of how a spinal motoneuron might integrate information from several sources.
10. Give examples of how levels of consciousness, language, and memory can be studied scientifically and related to cerebral function.
11. Explain the terms in italics in the following passage from the instructor in a childbirth preparation class: "During labor and delivery, a general anesthetic reduces the level of consciousness by affecting *synaptic transmission* in the CNS, particularly in the *reticular formation;* on the other hand, a local anesthetic can be injected to block *conduction* in the *spinal cord, roots* or *peripheral nerves* at *a sacral level.*"

2. The Nerve Impulse

How nerve cells become excited and conduct impulses

The nerve cell is the building block of the nervous system, and its role is to carry signals. The first signal to consider is the nerve impulse and its electrical manifestation, the *action potential*. The action potential, like the other signals of nerve cells, depends on electrical and chemical changes across the nerve membranes. Understanding these electrical-chemical changes may be difficult, but is important in interpreting (1) normal nerve activity, (2) abnormal nerve activity (as in epilepsy or multiple sclerosis), and (3) the action of drugs (such as local anesthetics). The same concepts are useful in the study of physiology of heart and skeletal muscle.

To begin with, consider a single neuron; for example, a motoneuron, which has its cell body in the spinal cord and sends its axon via the sciatic nerve to the leg (figure 2–1). The axon of this neuron, like other axons or nerve fibers, is specialized to communicate information from one place to another—in this case, from the spinal cord to the lower leg, a distance of about one meter. This information is sent by means of nerve impulses, which can be electrically recorded as action potentials. Nerve impulses are electrochemical changes across the nerve membrane. These changes affect only a small section of the axon at any instant but sweep down the axon from one end to another at rates of up to 200 mph. When the nerve membrane is in the resting state, an electrical voltage difference, called the *resting potential,* exists across the nerve membrane. The action potential is a sudden, large electrical change from this resting potential.

The Nerve Membrane

The nerve membrane (figure 2–2) separates the internal solution (axoplasm) from the external solution (extracellular fluid). According to classic concepts, it is composed of a *lipid bilayer,* in which the nonpolar chains face toward the center of the membrane and the polar ends face outward. On the sides of the membrane are molecules of *protein.* There are thought to be aqueous *channels* or *pores,* lined with protein, through which water and certain small ions can travel. Since only certain ions can pass through it, the membrane is called *semipermeable.*

The solutions inside and outside the cell contain about 155 mEq/liter of cations (positively charged ions) and the same concentration of anions (negatively charged ions). However, their composition differs.

There is a high concentration of potassium ions (associated with large, organic anions) inside and a high concentration of sodium ions (associated with chloride ions) outside. We will see that this unequal distribution of ionic species leads to an electrical voltage (the resting potential) across the membrane, and that a change in this voltage gives rise to the nerve impulse.

The unequal distribution of sodium and potassium ions is maintained by a Na^+–K^+ pump in the membrane. This pump uses

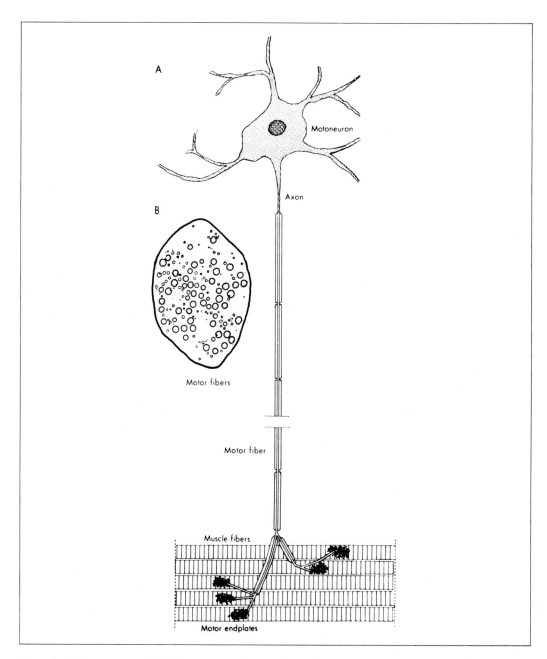

Figure 2–1. A motoneuron. (A) Motoneuron cell body, axon (nerve fiber), and terminals (motor endplates) on muscle fibers. (B) A peripheral nerve in cross-section, showing individual motor axons (the afferent fibers have degenerated). From Eccles, J.C. The Understanding of the Brain. New York: McGraw-Hill Book Co., 1973. Reprinted with permission.

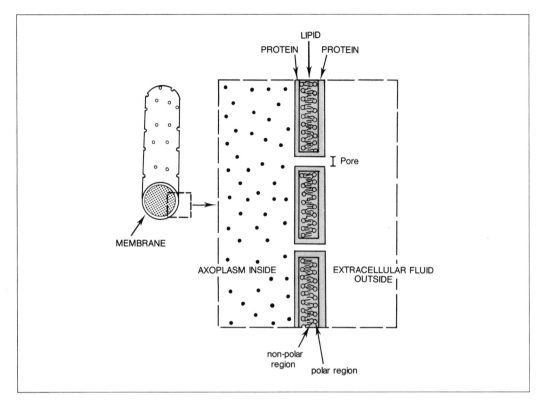

Figure 2–2. An unmyelinated axon (left) and class-ical model of its cell membrane (right) shown in schematic diagram.

Figure 2–3. The sodium–potassium pump in the nerve cell membrane.

metabolic energy to pump out any sodium ion that diffuses in, and, to a lesser extent, to pump back in the potassium ion that diffuses out (figure 2–3).

The excess sodium ion outside the nerve cell is also kept there because the sodium permeability of the membrane is very small in the resting state. The large anions (amino acids, protein, organic phosphate, sulfate) remain inside because they cannot permeate the membrane at all.

In the axon, 20% or more of all metabolic energy is used by the pump for active transport of sodium ions. The immediate source of this energy is adenosine triphosphate (ATP). One

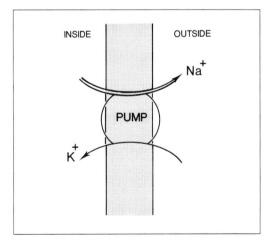

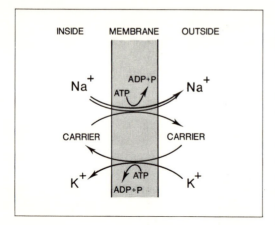

Figure 2–4. One model of the coupled sodium–potassium pump; the pump utilizes energy provided by the breakdown of ATP.

model for the pump is a carrier that transports sodium ions from inside to outside the cell and then combines with potassium ions and moves from outside to inside (figure 2–4). ATPase enzymes, which are activated by sodium and potassium ions, have been isolated in cell membranes and may be involved in this transport process.

Now that we know how the unequal concentrations of ions are maintained across the membrane, we can describe them in more detail and see how they produce a resting potential.

The Resting Potential

Dependence on K^+ Concentration Difference

Inside the nerve cell in the neuronal cytoplasm, the potassium ion is highly concentrated at 150 mM (all concentrations are approximate only). These potassium ions are associated with equal concentration of large anions symbolized by A^-. Outside the nerve cell, in the extracellular fluid, potassium ion concentration is low (5 mM). Sodium chloride concentration is low inside and high outside the membrane (figure 2–5).

If the membrane were permeable only to potassium ions, potassium ions would diffuse through channels in the membrane from inside to outside, down their concentration gradient. They would leave behind the large anions, which cannot permeate the membrane. Since originally both internal and external solutions are electrically neutral (each has the same concentration of anions and cations), the slight excess of potassium ions that develops outside and the slight excess of anions that develops inside cause a potential difference across the membrane. At equilibrium, this potential difference exactly opposes further diffusion of potassium ions outward (positive repels positive). The concentration difference of potassium ions is balanced by an opposing potential difference. The potential difference can then be calculated from the concentration of potassium ions on the inside, or K^+_i, and the concentration of potassium ions on the outside, or K^+_o, according to this formula:

$$V_{inside} - V_{outside} = V = -61 \ \log \frac{K^+_i}{K^+_o}$$

$$= -61 \ \log \frac{150}{5}$$

$$= -61 \ \log 30$$

$$= -61 \times 1.47$$

$$= -90 \ mV^1 \ (inside \ negative \ to \ outside)$$

This potential difference is called the *Nernst equilibrium potential for potassium ions*, or E_K. Thus, normally:

$$E_K = -61 \ \log \frac{150}{5} = -61 \ \log 1.47$$

$$= -90 \ mV.$$

[1]$-90 \ mV = -\dfrac{90}{1000}$ volts or slightly less than 1/10 volt.

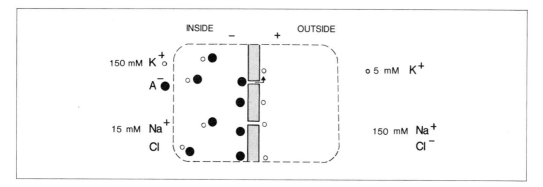

Figure 2–5. Diagram of the distribution across nerve membrane of ionic species that contribute to the resting potential. Concentrations are approximate only, and only a few ions are shown.

Changes in K^+_i or K^+_o lead to predictable change in E_K. Increasing K^+_o to 15 mM, for example, would mean that

$$E_K = -61 \log \frac{150}{15} = -61 \log 10 = -61 \text{ mV}.$$

Electrical Properties of the Membrane

We have already mentioned voltage, but other basic electrical concepts are also important in understanding what happens across the membrane during neural activity. We will start by describing a simple circuit—that of a flashlight. A simple flashlight circuit involves a battery, a conducting path (wire), and a light bulb (figure 2–6A). Positive charge flows from positive to negative poles of the battery, constituting a current (this is equivalent to saying that negative charge flows from negative to positive poles). The wire conductor can be cut and both cut ends inserted into an ionic solution (figure 2–6B). Current will continue to flow through the solution, carried by the dissolved ions: this flow is called *volume conduction*. A circuit diagram represents this flow (figure 2–6C), and the standard symbols are given in table 2–1.

Table 2–1. Common Electrical Symbols

Symbol	Term	Common units[a]	Meaning
V or E	Voltage or electrical potential difference	Volts (V)	The amount of electrical energy that potentially can be exerted on a charged particle such as an electron or an ion
I	Current	Amps (A)	The flow of charged particles through a conductive medium such as a metal wire or an electrolyte solution
R	Resistance	Ohms (Ω)	The resistance to the flow of charged particles through a conductive medium or across a membrane
G or g	Conductance	Mhos	The ease with which charged particles can flow through a conductive medium; the inverse of resistance (G = 1/R)[b]
C	Capacitance	Farads (F)	The ability of two conductors in the same vicinity to store charged particles on both sides of an insulating medium such as air or the non-porous area of a membrane

[a]Prefixes to units include: μ or micro-, one millionth; m or milli-, one thousanth; K or Kil(o)-, one thousand, M or meg(a)-, one million.
[b]The ability of a membrane to allow K ions to cross it is its potassium conductance, G_K or g_K, which is directly related to its potassium permeability, P_K; and similarly for other ions

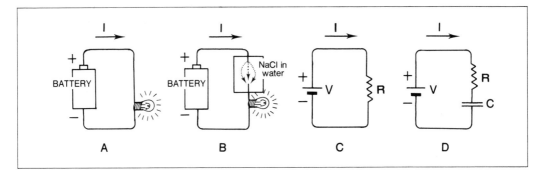

Figure 2–6. Simple electrical circuits. (A) Battery and light bulb, connected with wire conductor. Current (I) is shown as the flow of positive charge; this is identical to negative charge flowing in the opposite direction. (B) Same as A, except that current flows through a chamber filled with salt solution. (C) Schematic circuit diagram of B, showing a voltage source (V) and a resistance (R). (D) Same as C, with the addition of a capacitor (C).

Electric current can be compared with the flow of water: the electrical potential (or voltage) difference produced by the battery can be compared to the pressure difference produced by a pump, the electrical conductor to a pipe, and the electric current in the conductor to the flow of water through the pipe. Electrical resistance is increased by such factors as a thinner wire conductor or a more dilute ionic solution, just as resistance in a hydraulic system is increased by a narrower pipe.

The amount of current, I, flowing through the electrical circuit is directly proportional to the potential difference, V, and inversely proportional to the resistance, R; that is, $I = V/R$ (Ohm's Law). Conductance, symbolized by G or g, is simply the inverse of resistance ($G = 1/R$); a concentrated salt solution in figure 2–6B might have a low resistance and a high conductance, meaning that it conducted ionic current easily.

Finally, a capacitor—typically two thin metal plates separated by a nonconducting gap or sheet of paper—can be inserted into the circuit (figure 2–6D). Then current flows from

the battery charge only until the capacitor is completely charged with positive charge on one side and negative charge on the other. The capacitor is then able to store the charge on its plates, separated by the nonconducting element in between.

These symbols allow us to draw an electrical model of a segment of nerve membrane (figure 2–7). In the model, the concentration difference of K^+ across the membrane is the source of electrical pressure, which can be represented by a battery whose voltage is E_K, the Nernst equilibrium potential for K^+ ($E_K = -90$ mV).

Figure 2–7. Electrical model of a segment of resting nerve membrane. V_m is the voltage (potential) difference across the membrane, or $V_{inside} - V_{outside}$. E_K is the sodium equilibrium potential, which is the source of the voltage difference. G_K is the potassium conductance, which can be attributed to membrane channels that allow potassium ions to pass through. C_m is the membrane capacitance, attributed to the lipid layers of the membrane which separate ions.

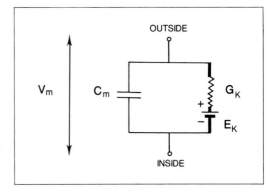

This concentration difference, represented by E_K, can be visualized as pushing potassium ions out through channels through the membrane. These channels offer a resistance, R, or, looked at in a more positive light, provide a conductance, G_K, to the flow of potassium ions. The potassium ions that pass through the potassium channels may be thought of as huddling against the outside wall of the nerve membrane, attracted there by the large negative ions they have left behind on the inside. The ability of the membrane to hold opposite charges across it is its capacitance, C_m, and is a property of the lipid layers that do not allow ions to cross. Once a certain number of potassium ions have collected on the outside of the nerve membrane, they repel the outward movement of their positively charged brethren because of simple electrostatic principles; this standoff represents an equilibrium situation, which to a large extent determines the resting potential. The effect of the potassium ion concentration difference is to deposit positive charges on the outside surface of the membrane capacitance and to leave the inside negatively charged. Only a few charges, charging the membrane in this way, are enough to cause a resting potential; they are too few to change the concentration difference across the membrane.

Actually, the membrane is somewhat permeable to other ions as well as K^+. Thus, the membrane potential at rest is not exactly equal to E_K (−90 mV); it is somewhat less negative (−70 mV to −85 mV).[2] This electrical voltage

[2]The contribution of the other ions to the membrane potential is expressed by the Goldman equation:

$$V_m = -61 \log \frac{P_K\,(K^+_i) + P_{Na}(Na^+_i) + P_{Cl}(Cl^-_o)}{P_K(K^+_o) + P_{Na}(Na^+_o) + P_{Cl}(Cl^-_i)}$$

P_K, P_{Na}, and P_{Cl} are the membrane permeabilities of the different ions. The contribution of each ion to this equation depends on its permeability. A potassium permeability much greater than that of the other ions would result in a membrane potential close to the Nernst equilibrium potential for potassium.

or potential during the resting state is called the resting potential. As we shall see, the resting potential is a baseline from which various changes—including the action potential—take off.

Changes in Membrane Potential

Changes in membrane potential (from the resting-potential level) can be measured by inserting small-tipped electrodes into the cell. These *microelectrodes* are made from thin glass capillary tubing. The tubing is heated and pulled to form a micropipette or microelectrode with an extremely small but open tip. The microelectrode is filled with an ionic solution and connected to a voltage-recording device to measure the voltage in a neuron. The recording device measures the membrane potential, in millivolts, between the microelectrode inside the cell and a reference electrode outside the cell in the extracellular fluid (figure 2–8).

Originally, when both microelectrode tip and reference electrode are outside the nerve fiber, they record the same electrical potential, so the potential difference between them is zero. Once the microelectrode pierces the nerve membrane and enters the axoplasm, the recorded potential difference goes suddenly negative to about −70 mV, the resting potential. In other words, the resting membrane is *polarized* to about −70 mV, inside negative. Now, electrical stimuli can cause the membrane voltage to change slightly, so that it is either slightly *depolarized* (less negative on the inside, for example, −65 mV); or *hyperpolarized* (more negative on the inside, for example, −80 mV). A weak stimulus from an electrical stimulator, as shown in figure 2–9, produces a slight depolarization in the membrane potential, as registered by a nearby microelectrode. Other microelectrodes 1 mm and 2 mm away record similar but much reduced voltage changes: the depolarization decays with distance (figure 2–9). A little farther along the

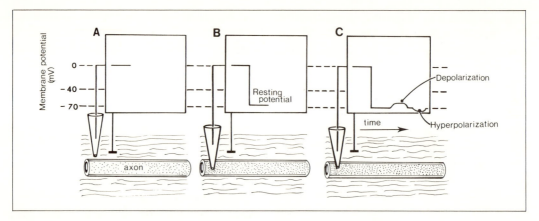

Figure 2–8. Membrane potential measured in a large nerve axon. Graphs show potential versus time as it would appear on an oscilloscope screen. (A) Microelectrode outside axon; the recorded potential difference = 0; (B) Microelectrode inside axon; the recorded potential difference equals membrane potential of –70 mV, the resting potential. (C) Slight (subthreshold) depolarization and hyperpolarization of the membrane from the resting level.

axon the depolarization disappears completely. However, signals in the human nervous system must be transmitted for distances up to almost 2 m, which is done by means of the action potential.

Figure 2–9. A subthreshold depolarization, showing its decay with distance. The electrodes are about 1 mm apart, and all traces start at the time the stimulus (S) is applied (arrow).

The Action Potential

The action potential, as shown in figure 2–10, begins when a depolarization goes beyond a certain *threshold* value—usually about 15 mV from the resting potential (for example, from –70 mV to –55 mV). A rapid depolarization follows and continues until membrane polarity is actually reversed; inside becomes briefly positive to outside, overshooting the zero mV line by about 30 mV to 40 mV. This is followed by a rapid repolarization back to resting levels. The whole process is called an action potential; due to its appearance, it may be also called a spike.

The action potential is conducted along the axon and does not decay with distance. It does not change in either height or time-course

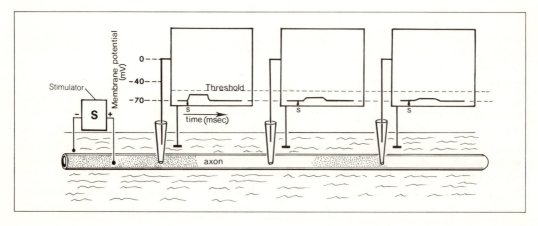

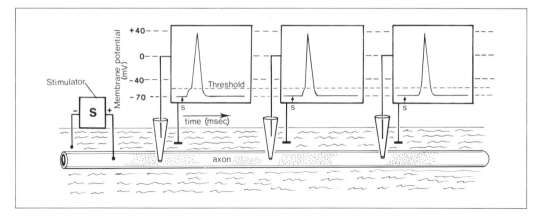

Figure 2–10. An action potential, originating with a stimulus applied near the left end of the axon. The stimulus is stronger, but otherwise the experimental preparation is the same as in figure 2–9. Note the absence of decay with distance.

$$E_{Na} = -61 \log \frac{Na^+_i}{Na^+_o} = -61 \log \frac{15 \text{ mM}}{150 \text{ mM}}$$

$$= +61 \text{ mV (inside positive)}$$

when recorded some distance away; this makes it a good signal for transmitting information. Also, as long as the depolarizing stimulus is over a threshold value, further changes in the stimulus have no effect on the height or shape of the action potential. This characteristic is known as the *all-or-none law:* the action potential is either full size or it does not occur at all.

The Ionic Basis for the Action Potential

The action potential occurs because, at threshold depolarization, Na^+ permeability channels in the membrane suddenly open, and the membrane quickly becomes much more permeable to Na^+. The sodium permeability, P_{Na}, and thus the sodium conductance, G_{Na}, increase rapidly. Na^+ then rushes into the cell, driven by its concentration gradient (Na^+_o is greater than Na^+_i) and by the voltage difference across the membrane. If Na^+ concentration alone were to determine membrane potential, this potential would be given by the Nernst equation:

The voltage never actually reaches this point but stops at about $+30$ mV to $+40$ mV when repolarizing processes take over.

In the electrical model, the increase in G_{Na} during the depolarizing phase of the spike can be represented by adding another section (figure 2–11). The sodium section is drawn heavier than the potassium section because the conductance to sodium ion is (briefly) much greater than the conductance to potassium ion during the rising phase of the spike. (While at rest $G_{Na} =$ about $1/20$ G_K; at the peak of the action potential G_{Na} has increased more than 100 times its resting value in a very short time).

Changes in ionic conductance can be graphed on the same time axis as the action potential (figure 2–12):

1. Sodium conductance rises quickly, causing rapid *depolarization.*
2. At the peak of the action potential, the inside becomes positive—the degree to which it becomes positive is called the *overshoot.*
3. Only a very small number of Na^+ ions move inward during each action potential (about

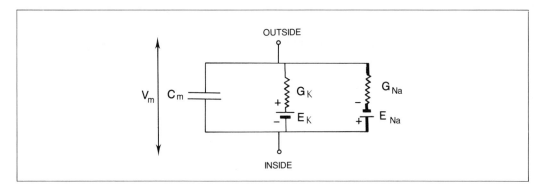

Figure 2–11. Electrical model of a segment of nerve membrane during the depolarization phase of an action potential. Identical to model of resting membrane (figure 2–7), with the addition of a significant sodium conductance (G_{Na}), due to the opening of sodium channels, which allows the sodium equilibrium potential (E_{Na}) to contribute a large part of the membrane potential. Positive ions pass inward through these channels and are separated from negative charges on the outside by the membrane capacitance. The result is a reversal of membrane polarity.

10^{-12} mole/cm^2 of membrane). Although sufficient to reverse the membrane potential, this ionic movement is not sufficient to significantly change the concentration on both sides. The excess Na$^+$ ions are gradually pumped out.

4. Near the action potential peak, sodium channels close down again (sodium conductance declines), while potassium channels open more than in the resting state (potassium conductance rises). Both factors cause *repolarization* to the resting level and also a period of slight *hyperpolarization* (the hyperpolarization occurs because the increased potassium conductance brings the membrane closer to E_K (–90 mV) than during rest).

After the onset of the spike there is an *absolute refractory period* in which a second spike cannot occur, no matter how strong the stimulus. This is followed by a *relative refractory period*, in which a second spike can occur but

only if the second stimulus is unusually strong. A main cause of refractoriness is that, once the sodium channels have opened and then closed, it is difficult to get them to open again until a

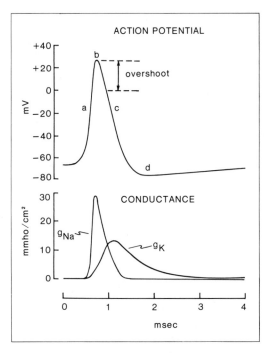

Figure 2–12. Conductance changes for sodium (g_{Na}) and potassium (g_K) drawn on the same time scale as the action potential. In the action potential trace, a = depolarization, b = peak of action potential, c = repolarization, and d = hyperpolarization. Source: Adapted from A.L. Hodgkin. The Conduction of the Nervous Impulse. Springfield, Ill.: Charles C Thomas, 1964.

few milliseconds have passed (the sodium conductance mechanism becomes inactivated).

The channels through which ions move
across the membrane have recently been the
focus of much attention [14,21]. These channels are thought to consist of proteins extending from one side of the membrane to the
other. Treating the membrane with certain
drugs and enzymes has shown that there are
separate channels for Na^+ and for K^+. Let us
consider the Na^+ channel. During the resting
state, the channel is effectively blocked by a
closed, protein gate structure. Depolarization
apparently causes a rearrangement within the
membrane of charged or dipolar particles associated with these gates (for example, the
positively charged groups of protein molecules
might be driven away from the channel, opening it to the passage of the positive sodium
ions). This rearrangement of charge within the
membrane produces a minute but measurable
current called the *gating current* and is associated with opening the gate to sodium. Thus
the tiny gating current is followed by a large
ionic current due to the sodium influx across
the membrane.

The different ion channels can be selectively
blocked by certain drugs. For example, *tetrodotoxin,* a poison obtained from fish, blocks
the sodium channels that allow sodium to move
inward during depolarization, and thereby pre-
vents the generation of an action potential.
Similarly, *tetraethylammonium* blocks the potassium channels. In addition, an inward movement of calcium through calcium channels
contributes in varying amounts to the action
potential in nerve and muscle cells, and certain
drugs have been found to block these channels.

Conduction of the Action Potential

At some segment of axon membrane, depolarization to threshold voltage can occur with
adequate stimulation. This segment generates
an action potential, marked by a brief reversal

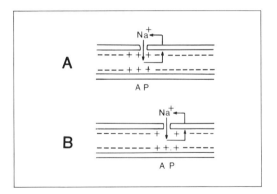

*Figure 2–13. Conduction of the action potential by
local circuits in an unmyelinated axon. (A) The
action potential (AP) has depolarized a section of
axon, while the section to its right is still in a resting
state (inside negative). The arrows show local circuits of ionic current. (B) An instant later, the section to the right has been depolarized and is now
generating an action potential, while the previously
active section to its left has become repolarized.*

in voltage difference across the membrane
(figure 2–13). In *unmyelinated axons* this active
region acts to depolarize the next adjacent
segment of membrane by means of local circuits. Local circuits are simply ionic currents in
extracellular fluid, axoplasnm, and membrane.
If they are blocked—by suspending a short
length of axon in air—conduction will cease.

In *myelinated axons* (figure 2–14) local circuits can only travel across the membrane at
the nodes between the myelinated, low-
conductance internodes. This makes for faster
conduction as each small segment of membrane
does not have to undergo the process of

*Figure 2–14. Saltatory conduction in a myelinated
axon.*

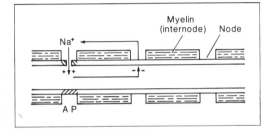

depolarization and excitation. Since the current jumps from one node to the next, conduction in myelinated axons is called *saltatory*, after the Latin for jump, *saltare*.

Demyelination occurs in multiple sclerosis and other diseases. The myelin sheath around the internode degenerates, so that some of the stimulating current passes through the internode and only the current left over passes through the next node. The current left over may be so small that conduction may be slowed or blocked. Ordinarily, the stimulating current passing through the node is more than enough to depolarize the node to threshold and excite an action potential. That is, the ratio of stimulating current to threshold current, called the *safety factor*, is greater than one. In demyelination, it may be reduced to less than one.

In both unmyelinated and myelinated nerve fibers, currents must travel longitudinally along the inside of the axon. The larger the axon diameter, the lower the internal longitudinal resistance, the greater the current flow, and the higher the conduction velocity. An additional factor increasing conduction velocity in myelinated fibers is the greater spacing between nodes found in larger-diameter axons. Experiments show that in myelinated axons the conduction velocity (in m/sec) increases with diameter, and equals approximately 6 × the diameter (in μ). For example, in myelinated fibers; an axon 2 μ in diameter conducts at 6 × 2 = 12 m/sec; an axon 5 μ in diameter conducts at 6 × 5 = 30 m/sec; and an axon 10 μ in diameter conducts at 6 × 10 = 60 m/sec. This relationship is graphed in figure 2–15. We will see in chapter 5 (Somatic Sensation) that fibers of different size and velocity have different functions. In addition, the conduction velocity of fibers in the human body can be measured, and is used as an index of nerve damage and recovery.

Figure 2–15. Conduction velocity of myelinated axons in meters per second (M.P.S.) plotted against axon diameter. From Gasser, H.S. Ohio Journal of Science 41:145–159, 1941. Reprinted with permission.

Recording of the Action Potential

We have described how action potentials are recorded with a microelectrode inside the nerve

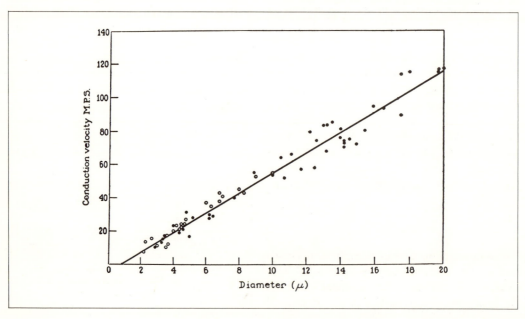

cell membrane. If the electrode is internal to a single neuron, it records a positive-going spike (figure 2–16A). However, if the electrode is external to a single neuron, it records a negative-going spike (figure 2–16B), since the outside of the membrane briefly becomes more negative during the spike. This record is obtained when the reference electrode is placed over a damaged section of nerve that does not conduct an action potential itself.

If the electrodes are placed outside a whole, compound nerve, consisting of many individual axons, a *compound action potential* is obtained (figure 2–16C). The compound action potential represents a combination of the voltage changes due to conduction in a large number of axons at the same time. As more fibers become active (with stronger stimulation) the response becomes larger. The response has several peaks representing fibers that conduct at different speeds. The first peak represents the largest-diameter, fastest-conducting fibers; the next peak represents smaller-diameter, slower-conducting fibers; and so on. Since the compound action potential can change in height and form, it does not look like an action potential from a single neuron and does not obey the all-or-none law.

Figure 2–16. Types of action potential recordings. (A) Intracellular (electrode tip inside nerve cell). (B) Extracellular (electrode tip outside nerve cell). (C) Compound action potential (electrode placed outside compound nerve). In B and C, nerve underneath reference electrode (rectangular plate) has been crushed so that it does not also generate an action potential, which would make recording more difficult to interpret.

Review Exercises

1. Describe the sodium–potassium pump in the nerve membrane.
2. State concentrations for sodium, potassium, and chloride ions within mammalian nerve cells and in the extracellular fluid.
3. Calculate the equilibrium potential for

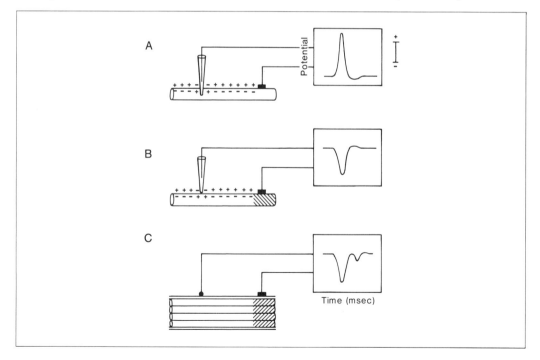

potassium and sodium using the Nernst equation.

4. Define and give units of measurement for electrical potential, resistance, current, capacitance.

5. Define conductance of the membrane and relate this to membrane resistance and ionic permeability.

6. Describe the technique for measuring the membrane potential of a neuron.

7. Define resting potential and state a typical value for this in a neuron.

8. Define depolarization, repolarization, and hyperpolarization.

9. Describe the spatial distribution and time-course of subthreshold changes in membrane potential.

10. Draw a diagram to show the time-course of the membrane potential changes in an axon when a sufficient stimulus is applied to elicit an action potential.

11. On such a diagram mark the depolarization phase, the spike potential, the overshoot, and the repolarization phase.

12. Describe the sequence of changes in membrane conductance (or permeability) to sodium and potassium ions that occurs on excitation of a neuron.

13. Describe the concept of sodium and potassium channels and the effect of tetrodotoxin.

14. When the fluid bathing an isolated axon was changed so that one-half of the external Na^+ was replaced with the larger, organic ion choline, the size of the overshoot was reduced. Explain why this occurred.

15. Describe the absolute and relative refractory periods.

16. "The absolute refractory period imposes an upper limit on the frequency of nerve impulses." Explain this statement as it would apply to a motor neuron with an absolute refractory period of 4 msec (1/250 sec). What would be the highest possible frequency of nerve impulses in spikes per second?

17. Draw a diagram to illustrate the local current flow that accompanies the passage of a nerve impulse along an axon.

18. Describe saltatory conduction.

19. In the Department of Physical Medicine, the conduction velocity of peripheral nerves is measured. A 28-year-old man was found to have a conduction velocity in the motor-nerve fibers of his arm of 31 m/sec, below the norm of 50 to 65 m/sec. Describe one possible cause of this showing.

20. Describe the relationship between the diameter of a myelinated nerve fiber and its conduction velocity.

21. Describe a compound action potential and how it differs from a single-axon action potential.

22. State the all-or-none law for nerve.

23. In the same patient described in question 19, sensory nerve impulses were studied by electrically stimulating the digital nerve in the middle finger with ring electrodes and a special stimulating circuit, while compound action potentials were recorded over the median nerve at the wrist. An intern assisting in this test noticed that as the size of the stimulus increased, so did the size of the recorded potential. He asked whether the equipment was faulty because he expected the all-or-none law to apply. Can you answer?

3. Sensory Reception

How receptors convert energy from the environment into nerve impulses

As you read this page, the words act on you by means of light energy. Music and speech act on you by means of sound energy. Another person's handshake presses on your hand, communicating by means of mechanical energy. Within the nervous system, information is transmitted by means of electrochemical signals called nerve impulses. Thus, the pressure from a handshake must somehow be converted into nerve impulses in the hand, which are then transmitted up the median, ulnar, and radial nerves in the arm to the spinal cord and, finally, to a sensory area of the cerebral cortex. How is the mechanical pressure first converted into nerve impulses in the hand? How do these nerve impulses carry information about the type and amount of pressure? How do similar processes occur for vision, hearing, and the other senses? These questions are taken up in this chapter.

Transduction in Mechanoreceptors

The process of converting one form of energy (in this case mechanical energy) into another form of energy (in this case the electrochemical energy of the nerve impulse) is called *transduction*. This transduction occurs in specialized cells in the skin and subcutaneous tissue called *receptors*, or more specifically, *mechanoreceptors*, since they respond best to mechanical energy.

Example 1: The Pacinian Corpuscle

Consider the pacinian corpuscle, a receptor found in subcutaneous tissue in the hand (figure 3–1). Under a microscope, it is an oval structure consisting of many layers (lamina), rather like an onion. Within the core of this structure is a nerve ending. While the ending of the nerve fiber is bare of myelin, a myelin sheath surrounds the fiber as it exits the corpuscle. The nerve fiber and the specialized tissue surrounding it together constitute a receptor. If you press quickly on someone's hand, the mechanical energy is transmitted through the skin to the pacinian corpuscles. What happens next was shown in a remarkable series of experiments using pacinian corpuscles isolated from animals, and poked by an electrically driven probe (figure 3–2A).

Receptor Potential. A mechanical indentation of the outer surface of the corpuscle is transmitted through the corpuscle to the membrane of the nerve ending at the core of the corpuscle. The nerve membrane converts or transduces the mechanical indentation into an electrical depolarization. In this depolarization, the inside of the nerve ending, at a resting potential of about -70 mV, becomes a few millivolts less negative, for a few milliseconds. This depolarization is called a *receptor potential* (because it takes place in a receptor) or a *generator potential* (because if large enough it can generate an action potential). The use of pharmacological substances that block action potentials has permitted the study of receptor potentials in isolation.

The receptor potential of the pacinian cor-

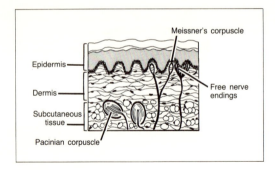

Figure 3–1. Mechanoreceptors in hairless skin, such as the palm or fingertips. Pacinian corpuscles, Meissner's corpuscles, and free nerve endings are shown.

puscle, like that of other mechanoreceptors studied, has these general characteristics:

1. Whether or not it generates an action potential, the receptor potential itself is a local depolarization, largely confined to the

Figure 3–2. Pacinian corpuscle and its axon. Receptor potentials and action potentials are recorded with one electrode on the axon and the other in the solution surrounding the corpuscle. **Th** *= threshold. Only response to stimulus onset, not offset, is shown.*

nerve ending and decaying rapidly with distance away from the point of stimulation. This spatial decay is like that described for subthreshold depolarization of a nerve axon in chapter 2 (see figure 2–9).

2. The receptor potential is graded in amplitude, in that a more intense stimulus produces a greater degree of depolarization, as shown (figures 3–2B and 3–2C). This applies to a wide range of stimulus strength, but at a certain point the degree of depolarization reaches a plateau, and further increases of stimulus intensity have no additional effect.

3. The receptor potential can be summated: if two weak stimuli are separated by only a short time interval, their receptor potentials will sum to produce a larger depolarization.

4. The receptor potential has no refractory period in that, unlike an action potential, one receptor potential can immediately follow another, with no delay.

5. The receptor potential appears to be due to increased ionic permeability to sodium ions and possibly other ions. Although the mech-

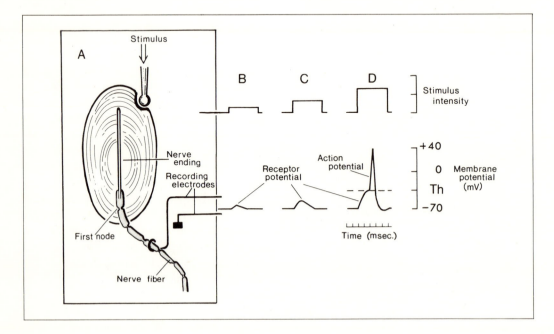

anism is not known definitely, the mechanical stimulus to the membrane may cause a movement of its protein components, opening ion channels and allowing increased ionic permeability.

Initiation of Action Potential. Although a very small stimulus will initiate only a small receptor potential with no action potential (figure 3–2B or 3–2C), a slight increase in the stimulus strength will cause a receptor potential large enough to cross threshold and trigger an action potential (figure 3–2D). The trigger point for the action potential—that is, the region with the highest sensitivity (lowest threshold) for starting an action potential—is the first node of Ranvier. Once begun at this point, the action potential or nerve impulse will be conducted along the nerve axon to the central nervous system.

What kind of signal does the nerve impulse provide the central nervous system? As you know, the amplitude and shape of the action potential do not change with stimulus intensity. All that can change is whether an action potential occurs in the first place and the timing and frequency of the nerve impulses as they follow one another. The occurrence of an action potential provides a signal that the stimulus is strong enough to cross threshold; the timing and frequency of the impulses depends on the stimulus and on the characteristics of the receptor, especially its adaptation.

Rapid Adaptation. The pacinian corpuscle is a rapidly adapting receptor. A quick mechanical stimulus above threshold causes a receptor potential leading to an immediate burst of one, two, or three nerve impulses. But the receptor potential quickly dies away; it rapidly adapts to the continued stimulus. As the receptor potential dies away, the nerve impulses also cease. Even if the indentation is maintained, there will be no more impulses during the indentation.

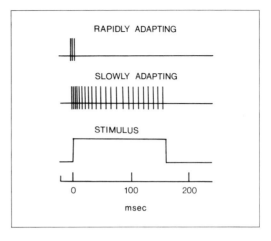

Figure 3–3. Rapidly adapting and slowly adapting receptors are compared. Only action potentials (spikes) of the receptors are shown, with an identical stimulus given to both types of receptor.

In contrast, there are other receptors that are slowly adapting, which continue firing nerve impulses for the duration of the stimulus. Both cases are illustrated in figure 3–3.

Thus the pacinian corpuscle is not well suited to transmit information about a continued mechanical stimulus. Instead, it is best suited to transmit information about a rapidly changing or *phasic* mechanical stimulus, which can be produced by a vibratory movement of an experimental probe. In everyday life, a vibratory stimulus can be produced by a tuning fork placed over the skin, by touching an automobile engine, or by moving the hand across a corrugated surface. If the strength of the vibration is high enough, each vibratory cycle triggers one or more nerve impulses, which are then transmitted from the pacinian corpuscle to the central nervous system (figure 3–4).

Other Rapidly Adapting (Phasic) Mechanoreceptors. The fingertips, obviously one of the areas of the body with the most acute sense of touch, contain other rapidly adapting mechanoreceptors in addition to the pacinian corpuscles. These receptors are called *Meissner's corpuscles*

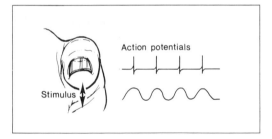

Figure 3–4. A vibratory stimulus to a fingertip, plotted against time (lower right), and the action potentials initiated in a pacinian corpuscle in response to that stimulus (upper right). The timing of the spikes reflects the timing of the stimulus.

(figure 3–1). Meissner's corpuscles are most sensitive to lower rates of vibration (e.g., about 30 Hz) while pacinian corpuscles are sensitive to higher rates (e.g., about 300 Hz). Thus, each type of receptor is tuned to a part of the spectrum of vibration and movement speeds found in nature.

On the back of the arm and on other areas of skin, hair-follicle receptors, with nerve endings coiled at the base of hair follicles, are also rapidly adapting and most sensitive to move-

ment. Bending the hair activates the receptor, but holding it in one position has no effect.

Example 2: The Stretch Receptor

The stretch receptors in lobsters and other crustaceans are easily studied because their large cell bodies are outside the central nervous system and send dendrites to adjacent muscle fibers. A microelectrode can be inserted into the cell body in order to record both the receptor potential generated in the dendrites and the action potentials generated in the axon (figure 3–5). This receptor transduces a mechanical change—stretch of the muscle—into a depolarizing receptor potential. As in the pacinian corpuscle, the receptor potential is (1) a local depolarization confined to the vicinity of the nerve ending without spreading significantly into the axon; (2) graded in amplitude with increased stimulus strength (as shown in figure 3–5); (3) capable of summating; (4) without a refractory period; and (5) due to increased ionic permeability. If the receptor potential is above threshold (figure 3–5D), action potentials will be generated at the initial segment of the axon (near the point at which it leaves the cell body), just as the pacinian corpuscle generates action potentials at its first node of Ranvier.

Unlike the pacinian corpuscle, however, the stretch receptor is *slowly adapting*. That is, during a maintained stimulus (a constant

*Figure 3–5. A crustacean stretch receptor stimulated by increasing its length by the amount Δ L. Both receptor and action potentials are recorded with a microelectrode in the cell body. **Th** = threshold. Compare with figure 3–2.*

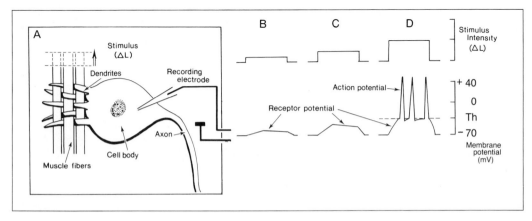

stretch) of adequate intensity, the frequency of the action potentials is greatest at the onset of the stretch, and there is then an adaptation (decline of the response) to a lower level, but this adaptation is slow and some response is maintained throughout the stimulus (figure 3–5D). Thus the stretch receptor is useful for detecting steady stretch of the muscle as well as changes in stretch. In general, slowly adapting (*tonic*) receptors are useful for detecting steady stimuli as well as changing stimuli, while rapidly adapting (*phasic*) receptors are useful only for detecting changing stimuli.

Like other sensory receptors, the crustacean stretch receptor signals the central nervous system by means of the timing and frequency of nerve impulses. In the stretch receptor, the onset of a series of impulses tells the nervous system approximately when a muscle stretch begins. During the stretch, the frequency of impulses (in spikes/sec) tells the nervous system the degree of stretch; a *greater stimulus intensity*, over a wide range, produces a greater receptor-potential amplitude, which produces a *higher spike frequency* (figure 3–6).

A higher spike frequency means that spikes come closer together in time, up to a limiting value. Certainly, the interval between two spikes cannot be shorter than the absolute refractory period. Thus if the absolute refrac-

tory period is 2 msec (1/500 sec), 500 spikes/sec is the absolute upper limit of spike frequency for that neuron.

Human skeletal muscles are supplied with stretch receptors that behave in many ways similar to those of crustaceans, but in human stretch receptors the cell bodies are distant from the sensory endings. The sensory nerve endings are attached to muscle fibers in such a way that a stretch of the muscle deforms the nerve endings. A depolarizing receptor potential follows, which, if above threshold, leads to a series of nerve impulses. The timing and frequency of the impulses inform the CNS about the change in length of the muscles, information used in controlling muscle contraction.

Other Slowly Adapting (Tonic) Mechanoreceptors. On the fingertips, there are slowly adapting receptors with disc-shaped endings (*Merkel's discs*) that transmit information about the degree of pressure exerted on the fingers. In the joint capsules, there are slowly adapting joint receptors that signal the angle of the limb at the joint, as well as the direction and rate of limb movement.

General Classification and Function of Sensory Receptors

There are other types of mechanoreceptors in addition to those described above. These include pain receptors (or *nociceptors*, from the

Figure 3–6. Response of a slowly adapting receptor (crustacean stretch receptor in this example) to increased stimulus intensity.

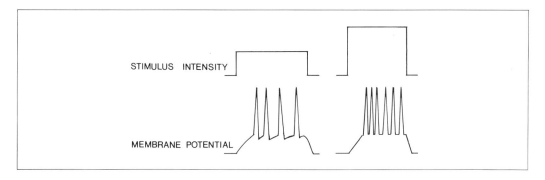

same root as the word noxious), which are specialized to transmit information about severe mechanical changes (cutting, crushing, and so on), or other stimuli that could cause damage. In the cochlea of the ear, specialized mechanoreceptors are sensitive to the minute air-pressure oscillations caused by sound waves. In the vestibular apparatus, other mechanoreceptors are sensitive to movements of the head and its position with respect to the pull of gravity.

In addition to the mechanoreceptors, there are receptors specialized to receive and transduce other forms of energy from the environment. *Thermoreceptors* are most sensitive to thermal energy (heat and cold). *Chemoreceptors* are most sensitive to various forms of chemical energy, for example, the solutions that cause taste sensations in the tongue and the gases that cause odor sensations in the nose. *Photoreceptors*—the rods and cones in the retina—are most sensitive to light energy.

Each receptor, then, is especially sensitive to a particular form of energy, which is called its *adequate stimulus* when the intensity is above threshold. The receptor transduces this form of energy into receptor potentials and then action potentials. The action potentials activate a sensory pathway in the central nervous system. Since the action potentials from different receptors are similar, it is believed that the *type (or modality) of sensation associated with each receptor is due to its central nervous system connections*—a modern restatement of one of the earliest generalizations in physiology, the law of specific nerve energies by Johannes Müller.

The importance of central nervous system connections allows us to understand why sensations can sometimes be produced by unusual stimuli. For example, rubbing the eyes (mechanical stimulation that may activate sensory neurons in the retina) may produce the illusion of lights or colors. Experimentally, electrical stimulation of auditory neurons can produce the illusion of sound. Electrical stimulation of the median nerve may produce the illusion that the hand is being touched. In certain diseases that affect the peripheral nerve, painful or tingling feelings may occur in an arm or leg without any outside stimulus. These *paresthesias* may be due to some abnormal stimulation of the peripheral nerve involved or of the CNS connections directly.

Review Exercises

1. Define transduction and explain its involvement in receptor function.
2. Describe the structure, electrical activity, and function of pacinian corpuscles, muscle stretch receptors, and two other examples of mechanoreceptors.
3. Define the receptor potential.
4. Compare the receptor potential with the action potential. Describe how receptor potentials can elicit action potentials.
5. Define adaptation as applied to receptors and contrast slowly adapting or tonic receptors with rapidly adapting or phasic receptors.
6. Cite examples of rapidly and slowly adapting receptors, and describe their functions.
7. Describe how the timing of afferent nerve impulses can be related to vibratory movement (for phasic receptors) and the intensity of a mechanical input (for tonic receptors).
8. Speculate about the effect on sensory experience of a mythical drug that temporarily makes all receptors rapidly adapting.
9. State and explain the significance of the law of specific nerve energies.
10. Make a list of receptors classified in terms of the form of energy they transduce.

4. Synaptic Transmission

How signals pass from neuron to neuron and from neuron to muscle fiber

In the last chapter we saw how a nerve impulse can travel the length of a nerve axon—even a distance as long as one meter as in the case of the motor–nerve fibers in the leg. What happens when the nerve impulse reaches the end of the nerve fiber? Since nerve and muscle fibers are separated by a gap, how does activity reach the muscle fibers to cause muscle contraction? This process, called *neuromuscular transmission,* is one form of *synaptic transmission,* the transmission of information from nerve cell to muscle, nerve, or gland cell. Not only is synaptic transmission a key to the complex behavior of the nervous system, but its disorder gives rise to major diseases (such as Parkinson's disease), and its modification is the basis of the action of many drugs (such as general anesthetics and tranquilizers) that affect the nervous system.

Long before the development of the neurosciences, a group of Indians living in the jungles of Ecuador modified neuromuscular transmission with a drug. Their survival depended on hunting tropical birds with a blowgun—a thin hollow reed about 10 feet long. The arrow that fit inside was formed from a palm leaf with a point as sharp as a needle. The arrow tip was dipped in a poison called *curare,* which was prepared from the wild Wourali vine. Even at distances of 200 feet, the hunters seldom missed the flying birds. Within three minutes of being hit by the arrow tip, the bird became motionless and fell dead. Curare is a neuromuscular blocking agent; it blocks neuromuscular transmission and thus paralyzes the muscles.

Similar drugs are now used in a controlled way to aid muscle relaxation during surgery. The effect of such drugs illustrates how the synaptic junctions between cells provide a site of action for chemicals. The chemicals can be *poisons,* as in the case of curare; *drugs,* as in the case of the blocking agents used during surgery; or substances normally found in the body, such as *chemical transmitters.*

Nerve–Muscle Synapses

Neuromuscular transmission is the transmission of action potentials from nerve fiber to muscle fiber by means of a chemical intermediary. A single motor-nerve fiber will branch and contact from 10 to 2,000 separate skeletal muscle fibers; since the nerve fiber triggers an impulse in all the muscle fibers it innervates, the nerve and muscle fibers together are called a *motor unit* (figure 4–1). Fibers of different units are interspersed with each other to ensure smooth contraction of the muscle as a whole.

The *neuromuscular junction* (figure 4–2) is where the nerve ending meets the muscle fiber. As it nears the muscle fiber, the motor axon loses its myelin covering and ends in a *presynaptic terminal.* In microscopic sections of the terminal one can see small *synaptic vesicles.* A *synaptic cleft* of about 50 nm (nanometers)[1] separates the nerve ending from the specialized *postsynaptic membrane* of the muscle fiber. In this region the muscle membrane is thrown

---

[1]One nanometer equals 10^{-9} meter or 10 angstrom units.

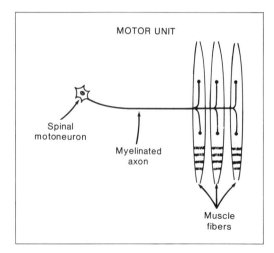

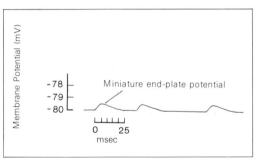

Figure 4–3. *Miniature end-plate potentials (MEPPs) recorded from muscle fiber at neuromuscular junction.*

Figure 4–1. A motor unit consisting of a spinal motoneuron with its myelinated axon innervating several skeletal muscle fibers.

into a series of folds (the junctional folds), which increase its surface area. The neuromuscular junction is also called the motor *end-plate*.

Neuromuscular Transmission

The synaptic vesicles resemble small membrane-bounded bubbles, about 50 nm in diame-

Figure 4–2. A neuromuscular junction (schematic diagram). Note the synaptic vesicles within the nerve ending (presynaptic terminal).

ter, within the cytoplasm of the nerve ending. The vesicles contain *acetylcholine*, a chemical transmitter. In brief, the role of the acetylcholine is to leave the vesicles, cross the synaptic cleft, contact the muscle membrane on the other side, and there cause electrical depolarization.

Even when muscle fibers are at rest, without the influence of nerve impulses, the effects of acetylcholine can be seen. The resting potential for muscle fibers is about −80 mV (as in nerve cells, inside negative to outside). Superimposed on this resting potential, however, are tiny depolarizations of about 0.4 mV—that is, the potential may briefly become less negative, going from −80 mV to −79.6 mV and back again (figure 4–3). These small depolarizations, called *miniature end-plate potentials*

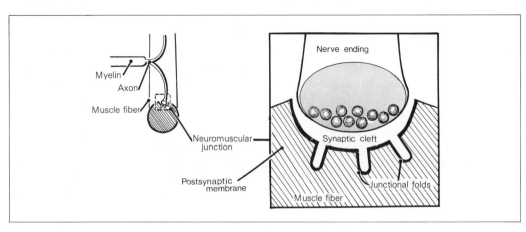

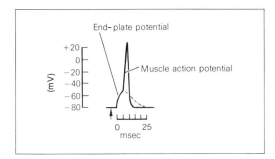

Figure 4–4. End-plate potential (EPP), leading to a muscle action potential. If action potential were not triggered at threshold, the EPP would continue as shown by dashed line before returning to resting potential. Arrow: arrival of impulse in motor nerve ending.

(MEPPs) occur randomly and apparently are due to the spontaneous release of acetylcholine from vesicles into the synaptic cleft. A single packet of acetylcholine, possible representing the contents of one vesicle, causes a depolarization of about 0.4 mV. These small depolarizations are far below the threshold value for the

Figure 4–5. Release of chemical transmitter (acetylcholine) from vesicles in the motor nerve ending. (A) The nerve action potential allows calcium ions to cross the presynaptic membrane and to enter the nerve ending. (B) Vesicles then fuse with presynaptic membrane, open, and release chemical transmitter contents into synaptic cleft. Chemical transmitter molecules then diffuse across to combine with receptor sites on postsynaptic membrane; the chemical fit between transmitter and receptor is shown schematically.

muscle fiber, so no muscle action potential occurs.

When an action potential arrives at the nerve ending, it causes the release of about 300 packets of acetylcholine, causing a full-sized *end-plate potential* (EPP, figure 4–4). The end-plate potential has a time-course similar to that of the miniature end-plate potential, but its much larger depolarization (over 30 mV) is enough to cross threshold and trigger a muscle action potential.

What events are behind the end-plate potential? When the action potential invades the nerve ending, it allows calcium ions to enter through the nerve membrane (figure 4–5). The calcium entry (possibly by changing the electrical charge on the vesicles) causes the vesicles to fuse with the nerve membrane, open, and release their contents, a process called *exocytosis*. The acetylcholine, once released, diffuses across the narrow synaptic cleft.

A *synaptic delay* of about 0.5 msec intervenes between depolarization of the axon terminal and the onset of the end-plate potential. This interval includes the time required for presynaptic release of the acetylcholine, its diffusion across the synaptic cleft, and its effects on the postsynaptic membrane.

At the postsynaptic membrane, the acetylcholine reacts with molecular sites called *acetylcholine receptors*. This reaction opens membrane channels to both sodium and potassium ions. Both ions then diffuse across the muscle membrane, seeking their electrochemical equilibria.

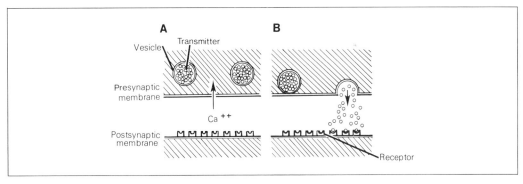

Using the Nernst equation, we can calculate that the equilibrium potential for sodium ion is positive ($E_{Na} = +61$ mV) while the equilibrium potential for potassium ion is negative ($E_K = -90$ mV). A compromise value between the two (($E_{Na} + E_K)/2$) is -15 mV. The end-plate potential is a depolarization that moves toward this value; however, an action potential

arises (and obscures the end-plate potential) after a smaller depolarization.

The location of the acetylcholine receptors can be shown by an experiment in which acetylcholine is released from a micropipette onto the muscle membrane (figure 4–6). When the micropipette is placed immediately over the neuromuscular junction, a normal-looking end-plate potential results. In this procedure, the micropipette is imitating the normal function of the nerve ending. However, if the micropipette is moved slightly to one side, the end-plate potential is much reduced. This illustrates that the acetylcholine receptor sites are concentrated at the muscle end-plate region.

The same acetylcholine receptor sites can be blocked by the poison *curare*. This blockade causes the end-plate potentials elicited by motor-nerve impulses to be much reduced in size, so that they fail to elicit muscle action potentials, and paralysis can ensue. A number

Figure 4–6. Experiment demonstrating concentration of acetylcholine receptors near motor nerve terminals. (I) Micropipette (E) releases acetylcholine at several points near nerve terminal. (II) Greatest sensitivity is at points a and d, adjacent to nerve terminals, and least sensitivity is at points in between. (III) Endplate potentials elicited at points a through d. Reprinted with permission from Eccles, J.C. The Understanding of the Brain. New York: McGraw-Hill Book Co., 1973. Adapted from Peper, K., and McMahan, U.J. Distribution of acetylcholine receptors in the vicinity of nerve terminals on skeletal muscles of the frog. Proc. Roy. Soc. Lond. (Series B) 181:431–440, 1973.

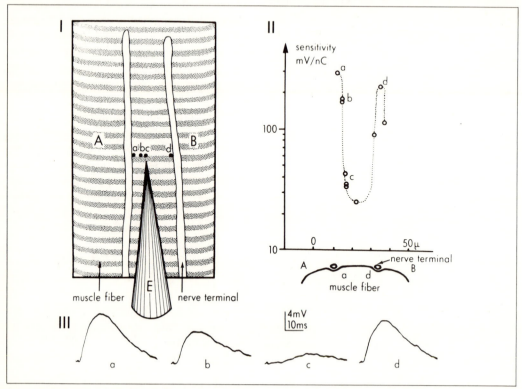

of medically useful drugs, such as *tubocurarine* (the main active constituent of curare), also block receptor sites; they cause skeletal muscle relaxation, which is helpful during many surgical procedures, and their harmful effects can be prevented by carefully monitoring the dosage and by providing artifical respiration if necessary. Another blocking agent and poison is *alpha-bungarotoxin,* which is extracted from the venom of *Bungarus* snakes. This toxin can be radioactively labelled, and after it binds onto acetylcholine receptors, its appearance in histological sections provides additional evidence that the acetylcholine receptors are concentrated at the end-plate region.

Neuron–Neuron Synapses

A *synapse* is the region at which one neuron relays messages to another. At a typical synapse (figure 4–7), the axon ending (or terminal) of one neuron ends close to the cell body, dendrite, or axon of another neuron. Chemical transmitters released by the axon terminal cross the synaptic cleft to bind onto the membrane of the second neuron, in a process very similar to neuromuscular transmission. The two major structures involved in a synapse are the presynaptic terminal and the postsynaptic membrane, separated by the synaptic cleft.

The *presynaptic terminal* is an expanded part of an axon, often at its ending but sometimes along its shaft. Because of their appearance, presynaptic terminals have also been called *presynaptic buttons* (or *boutons* in French). As many as 1,000 of these terminals, from numerous axons, may end on a single nerve cell. Presynaptic terminals contain synaptic vesicles similar to those in the motor-nerve endings at the neuromuscular junction. At some synapses, the vesicles contain acetylcholine. However, as Palay [25] pointed out, vesicles come, like chocolates, in a variety of shapes and sizes and are stuffed with various kinds of fillings. Their sizes range from 20 to 90 nm, and chemical transmitters include several substances in addition to acetylcholine.

The *synaptic cleft* is a space about 20 nm wide between the presynaptic and postsynaptic membranes.

The *postsynaptic membrane* is the specialized membrane of the nerve cell on the other side of the synaptic cleft. It often has a layer of dense material underlying it, and on dendrites, may be located on small knobs called *dendritic spines.*

Transmission at Excitatory Synapses

The physiological activity of a synapse can be illustrated by using a spinal motoneuron as an

Figure 4–7. A typical neuron–neuron synapse (schematic diagram).

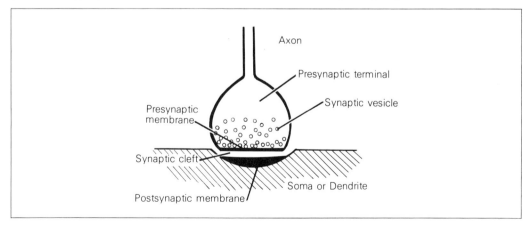

example (figure 4–8). In this experimental setup, a microelectrode is inserted into the cell body, and an oscilloscope shows the membrane potential—the voltage inside the cell minus the voltage outside it. At first, the membrane potential is at its resting level of –70 mV. At time zero, one or more presynaptic terminals become depolarized by an action potential. Although these inputs to the synapse may be either excitatory or inhibitory, we will begin by considering the events at excitatory synapses.

After the action potential invades the pre-synaptic terminal, the ensuing release of transmitter from the vesicles is similar to that in neuromuscular transmission (but the transmitter itself may be an amino acid or some other compound). Also similar is the synaptic delay of approximately 0.5 msec, the presence of receptor molecules on the postsynaptic membrane, and the effect of the transmitter reacting with the receptor molecules—an opening of membrane channels to both sodium and potassium ions, leading to a depolarization from the resting potential on the postsynaptic side.

Figure 4–8. Excitatory synaptic transmission, as shown by recording with microelectrode in spinal motoneuron. Inset shows superimposed EPSP responses to activation of 5, 25, or 50 excitatory synaptic inputs at time indicated by arrow. Once action potential begins at initial segment, it is conducted along axon. Reference electrode is outside the cell.

Excitatory Postsynaptic Potential. The postsynaptic depolarization is called an *excitatory postsynaptic potential* (*EPSP*). The diagram (figure 4–8) shows that five excitatory inputs cause an EPSP that depolarizes the membrane by about 5 mV, from –70 mV to –65 mV. In itself, such a small EPSP is a transient and

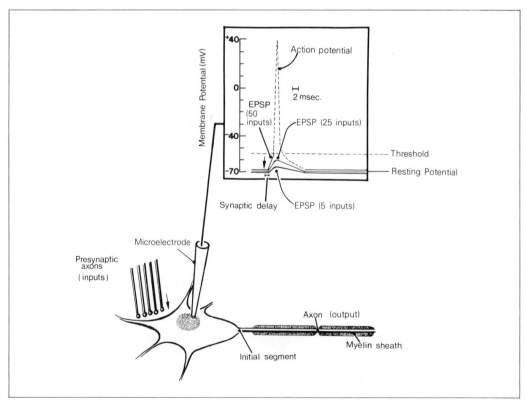

localized event; it is subthreshold and does not trigger an action potential. Instead, it dies away after about 15 msec, in a process called *temporal decay*. Also, even when the EPSP at the synapse is at its peak, it may be much reduced a few micrometers away from the synapse, a process called *spatial decay*. Thus, the EPSP by itself is a local process of limited extent, and, therefore, is of no use in transmitting a signal for any distance along the axon.

The amount of depolarization increases as more presynaptic fibers fire at the same time. The diagram shows that increasing the number of inputs from 5 to 25 increases the amount of depolarization to −60 mV. Thus, the EPSP is a graded process that can increase in size as a function of the amount of excitatory input.

Threshold. In the example shown in figure 4–8, only when 50 inputs fire does the depolarization reach the threshold value of −55 mV. At this point, an action potential starts in the initial segment, the most sensitive region of the motoneuron. The action potential than sweeps down the axon, jumping from node to node by saltatory conduction.

The number of inputs and the threshold voltage are mentioned only as typical values found in some spinal motoneurons. The general principle is that, in most neurons studied, more than one excitatory input is required to elicit an action potential output, and the action potential originates at a particular low-threshold area on the neuron, generally the initial segment.

Characteristics of Synaptic Transmission

The excitatory synapse on the spinal motoneuron exemplifies a number of general characteristics of synaptic transmission. These characteristics are found in other chemical synapses as well, including the neuromuscular junction and the inhibitory synapses, which remain to be considered.

One-Way Conduction. Synaptic transmission proceeds from presynaptic nerve terminals to the postsynaptic cell, rather than in the opposite direction. Although conduction of an action potential along an axon is also generally in one direction, it is possible for an action potential to move in both directions away from the site of stimulation if an axon is stimulated somewhere between the cell body and the axon terminal. Thus, synaptic transmission differs from action potential conduction in this respect and may be compared with one-way streets regulating the flow of automobile traffic during rush hour.

Synaptic Delay. The time between depolarization of the presynaptic terminal (generally because of an invading action potential) and the onset of depolarization in the postsynaptic membrane is approximately 0.5 msec. This is the time required for the release of transmitter, its diffusion across the synaptic cleft, and its activation of the postsynaptic membrane.

Graded Response. We have seen that the EPSP is graded in size, in that the amount of depolarization increases with the number of excitatory inputs active at the same time. The action potential, on the other hand, does not increase in size with increased input: it is an all-or-none event.

In neuromuscular transmission, a graded end-plate potential can be seen when the muscle action potential is blocked with curare. Under normal conditions, however, a nerve impulse in the motor axon results in an end-plate potential large enough to elicit a muscle action potential, so that gradation of response is not a factor.

Summation. The ability of the EPSP to increase when more and more excitatory synapses are simultaneously active is called *spatial summation*, since it represents a summing of inputs distributed along the spatial dimension of the postsynaptic neuron. In addition, when nerve impulses invade an axon terminal one after the other at a high enough rate, the resulting EPSP will also increase, a feature called *temporal summation.*

Under normal conditions, numerous synaptic inputs may arrive at different locations on the neuron, each at different frequencies and somewhat different times. Thus, both spatial and temporal summation play a part in determining whether or not an action potential is generated in the postsynaptic cell.

Transmission at Inhibitory Synapses

Some synaptic inputs have inhibitory instead of excitatory effects. In postsynaptic inhibition, the response of the postsynaptic membrane consists of an *inhibitory postsynaptic potential* (IPSP). In this case, the transmitter causes a simultaneous increase in membrane permeability to potassium and chloride ions. Since the equilibrium potential for potassium ion is −90

Figure 4–9. A typical IPSP in a spinal motoneuron, beginning at a −70 mV resting potential and moving in a hyperpolarizing direction. Inhibitory input occurs at time indicated by arrow.

mV, and that for chloride ion is −70 mV, the effect of the permeability increase for both ions is to drive the membrane toward the intermediate value of −80 mV. This represents a membrane potential that is more negative on the intracellular side (hyperpolarized) with respect to the resting potential of −70 mV.

When the membrane starts at its normal resting potential, the activation of an inhibitory synaptic input elicits an IPSP that consists of a hyperpolarization of a few millivolts or less, which resembles an inverted EPSP (figure 4–9). Like EPSPs, IPSPs have a synaptic delay, are graded according to the number of inputs activated, have no refractory period, show spatial and temporal summation, and decay over time and distance.

IPSPs have an effect antagonistic to that of EPSPs. While a large-enough EPSP will depolarize the membrane to the threshold level and trigger an action potential, a concurrent IPSP will tend to counteract this effect and keep the membrane potential below threshold so that no action potential results (figure 4–10).

An overall comparison of excitatory and inhibitory synaptic transmission is shown in figure 4–11. Both are initiated by action potentials (APs) arriving at presynaptic terminals. When the excitatory transmitter com-

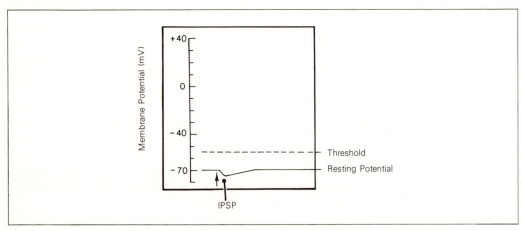

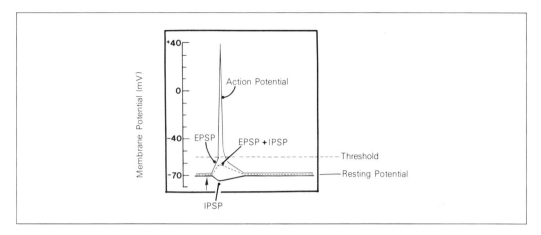

Figure 4–10. Above-threshold EPSP (upper trace), and IPSP (lower trace), triggered at time indicated by arrow. Dashed line: when IPSP is triggered simultaneously with EPSP, depolarization is below threshold.

bines with postsynaptic receptors, it produces a localized depolarization (the EPSP) via an increased permeability to sodium and potassium ions. In the postsynaptic neuron, the EPSP, if large enough to cross threshold, initiates an action potential that is then conducted along the axon as an output signal. A simultaneous IPSP, however, can prevent this by driving the membrane potential away from the threshold level.

*Electronic Logic Circuits
as Models for Synaptic Integration*

Electronic logic circuits are important components of electronic computers and illustrate some basic principles of synaptic integration (like any other model, however, their resemblance to the real thing is quite incomplete). Consider simple logic circuits with two inputs, A and B, and one output, C (figure 4–12). In the columns on the right side of the figure, the presence of a signal (equivalent to an action

potential) is indicated by a one, and its absence is indicated by a zero.

The columns under I illustrate the summation of two excitatory inputs to produce an action potential. An excitatory input signal in A alone or B alone does not produce an output signal in C; only when both excitatory inputs A and B are active (as shown in the bottom row) does an output signal, equivalent to an action potential, appear in C.

The columns under II illustrate the ability of an inhibitory input to prevent an output signal. An excitatory input in A alone, equivalent to a large EPSP, and the absence of an inhibitory input in B, produce an output signal in C (as shown in the second row). But an inhibitory input in B alone (third row) or together with an excitatory input in A (fourth row) produces no output signal in C.

Effect of Synaptic Location

The previous discussion has assumed nearly the same location for excitatory and inhibitory synapses on the postsynaptic neuron. In reality, synapses are located at quite different sites along the cell body and dendritic tree, and these locations affect synaptic interaction. For example, the model of a dendrite or dendritic

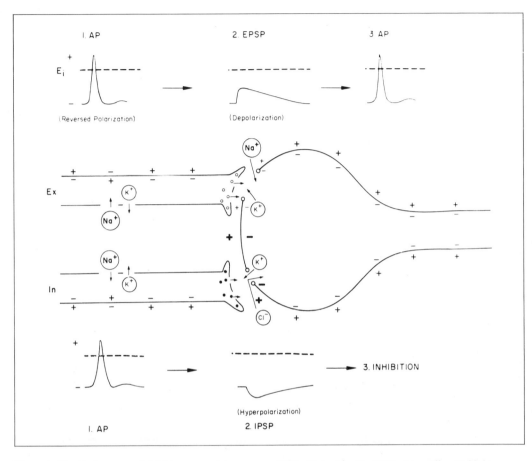

Figure 4–11. Excitatory and inhibitory synaptic transmission compared. (1) The action potential (**AP**), consisting of a reversal of membrane polarization, arrives at the presynaptic terminal and causes release of excitatory (upper half of diagram) or inhibitory (lower half) transmitter. E_i = internal potential; dashed line = zero potential; Ex = excitatory; In = inhibitory. (2) Combination of the excitatory transmitter with postsynaptic receptors produces a localized depolarization, the EPSP, through an increased permeability to sodium and potassium ions. The inhibitory transmitter produces a localized hyperpolarization, the IPSP, through an increased permeability to potassium and chloride ions. (3) The EPSP, if large enough (summation of several excitatory inputs is often necessary), initiates a conducted **Ap** in the postsynaptic neuron. However, this can be prevented by a concurrent IPSP. Reprinted with permission from Gilman, A.G., Goodman, L.S., and Gilman, A. The Pharmacological Basis of Therapeutics (6th ed.). New York: Macmillan, 1980. Copyright © 1980, Macmillan Publishing Co., Inc. Modified from Eccles, J. The Understanding of the Brain. New York: McGraw-Hill, 1973; and from Katz, B., Nerve, Muscle, and Synapse. New York: McGraw-Hill, 1976.

Figure 4–12. Electronic logic circuits as models of synaptic integration in a simplified neuron. Inputs **A** and **B** may be compared with presynaptic terminals, output **C** may be compared with axon. In tables on right, 1 indicates a signal and 0 indicates no signal.

		I			II		
		A	B	C	A	B	C
A	B	0	0	0	0	0	0
		1	0	0	1	0	1
		0	1	0	0	1	0
C		1	1	1	1	1	0

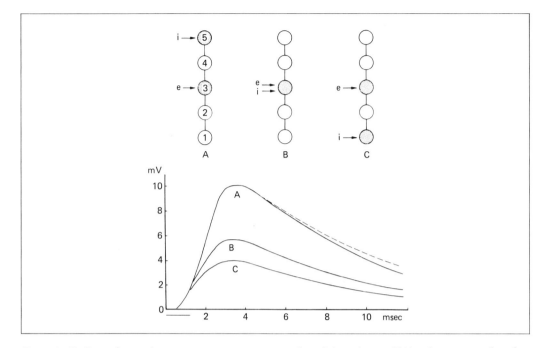

*Figure 4–13. Dependence of synaptic integration on location of synaptic inputs along a dendrite, after Rall [29]. Above: Numbered circles represent points along a dendrite, with 1 being at cell body and 5 being most distant. An excitatory synapse (**e**) is located in the middle in all cases, while an inhibitory synapse (**i**) is located more distantly (**A**), at the same site (**B**), or closer (**C**) to cell body. Below: Resulting synaptic potentials recorded at the cell body for these three cases. Dotted line represents response to excitatory synapse alone. Source: Shepherd, G.M. The Synaptic Organization of the Brain (2nd ed.). New York: Oxford University Press, 1979. Reprinted with permission.*

reduced by about 60% of its control value. Finally, the inhibitory input is most effective when located proximal to the excitatory input (C)—the recorded EPSP is now reduced to only 40% of its control value. The overall conclusion is that inhibitory synapses are more effective (with respect to control over the cell body) when they are closer to the cell body. The potential of the cell body, and the nearby initial segment, is important to the initiation of nerve impulses as output signals from the neuron.

tree shown in figure 4–13 has five possible synaptic sites, site 1 being closest to the cell body and site 5 being most distant. When excitatory synaptic inputs (indicated by e) and inhibitory synaptic inputs (indicated by i) occur at various combinations of sites, the computed synaptic potential (or EPSP) recorded at site 1 shows that the inhibitory input has little effect when significantly distal to the excitatory input (A). When both inputs are at the same middle site (B), the recorded EPSP is

Presynaptic Inhibition

Postsynaptic inhibition acts via an effect on the postsynaptic membrane (the IPSP). In presynaptic inhibition, on the other hand, the inhibitory neuron synapses onto the presynaptic terminal of an excitatory neuron (figure 4–14), and its effect is to reduce the amount of transmitter released by the presynaptic terminal with each nerve impulse. This reduces the size of the resulting EPSP without causing any

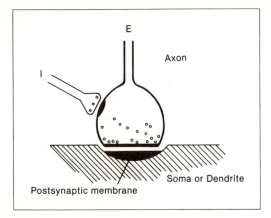

Figure 4–14. Presynaptic inhibition, involving axoaxonal synapses of inhibitory axon (I) onto excitatory axon terminal (E).

direct change of postsynaptic membrane conductance and also without affecting the response of the postsynaptic cell to other synaptic inputs. Thus, presynaptic inhibition may function in the CNS to reduce the effectiveness of particular synaptic inputs onto a neuron, without altering the effects of other inputs.

The structural basis of presynaptic inhibition is thought to be the synapse of an inhibitory axon (I in the diagram) onto an excitatory axon terminal (E), corresponding to the axoaxonal synapses observed under the electron microscope. The inhibitory axon releases a chemical transmitter that binds with receptor sites in the excitatory axon terminal. If the excitatory terminal is invaded by a nerve impulse within a short time interval it releases fewer vesicles of its chemical transmitter than it normally would.

Chemical Transmitters

Chemical transmitters are substances that are released by a presynaptic nerve ending after its stimulation, and cross the synaptic cleft to cause either excitation or inhibition in the postsynaptic cell. The best-known chemical transmitters are acetylcholine and norepinephrine (and its close relative epinephrine). A number of other compounds are presumed to be transmitters on the basis of several experimental criteria, but they are not yet accepted by all observers as proven transmitters; these presumed transmitters include dopamine, serotonin, and the amino acids GABA and glycine. More recently discovered are the enkephalins and endorphins, which are peptides that appear to attach themselves to the same receptor site as the drug morphine and may share some of its effects as a pain-relieving (analgesic) substance. The structures of these compounds are shown in figure 4–15.

Acetylcholine (ACh)

Acetylcholine is synthesized in neurons by a reaction between acetyl coenzyme A (CoA) and choline, catalyzed by the enzyme choline acetylase:

$$\text{Acetyl CoA} + \text{Choline} \xrightarrow[\text{acetylase}]{\text{choline}} \text{Acetylcholine}$$

The transmitter is stored in vesicles in the presynaptic endings and then released when an action potential sweeps over the endings (figure 4–16). After release, the transmitter diffuses across the synaptic cleft to combine with acetylcholine receptor sites on the postsynaptic side.

The transmitter released into the synaptic cleft is quickly removed or inactivated, allowing the transmitter secreted later to have a clear effect. In the case of acetylcholine, this removal and inactivation is accomplished by (1) diffusion of acetylcholine away from the synapse and (2) enzymatic breakdown (hydrolysis) into acetic acid and choline, under the influence of the enzyme acetylcholine esterase (cholinesterase). The products of the breakdown are then actively taken up by the presyn-

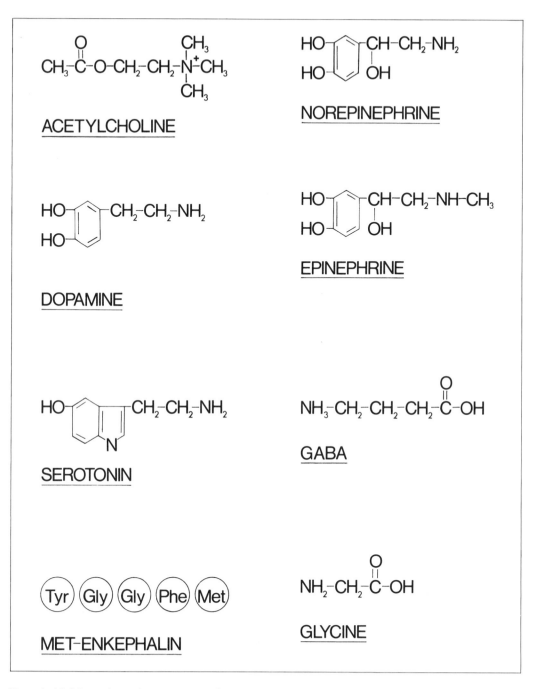

Figure 4–15. Major chemical transmitters and presumed transmitters.

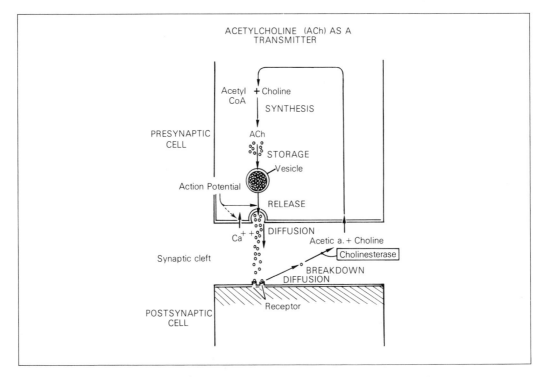

Figure 4–16. Acetylcholine (ACh) as a transmitter. Presynaptic action potential, in the presence of calcium, triggers release of transmitter, which diffuses across and combines with postsynaptic receptor sites.

aptic terminal and recycled to produce more acetylcholine.

Cholinergic neurons secrete acetylcholine as a transmitter; they include the following (figure 4–17):

1. Motoneurons (both in the spinal cord and in the cranial motor nuclei) and probably other neurons in the CNS
2. Preganglionic fibers of the autonomic system (both sympathetic and parasympathetic) and those that innervate the adrenal medulla (an organ with a similar embryologic origin to the sympathetic ganglia)
3. Postganglionic fibers of the parasympathetic system
4. Some specific postganglionic fibers of the sympathetic system (for example, *sympathetic cholinergic fibers*, which innervate sweat glands)

Norepinephrine (NE)

Norepinephrine (also called noradrenaline) is synthesized by a series of reactions involving several enzymes as shown in figure 4–18. Like acetylcholine in other nerve terminals, norepinephrine is stored in vesicles to be released during a nerve impulse (figure 4–19). Inactivation of the released transmitter (clearing the synapse so it can respond to the next impulse) occurs by means of (1) diffusion of norepinephrine away from the synapse; (2) enzymatic breakdown to an inactive metabolite, involving the enzyme catechol-O-methyl transferase (COMT); only a small amount of norepinephrine is inactivated by this pathway; and (3) active re-uptake by the presynaptic

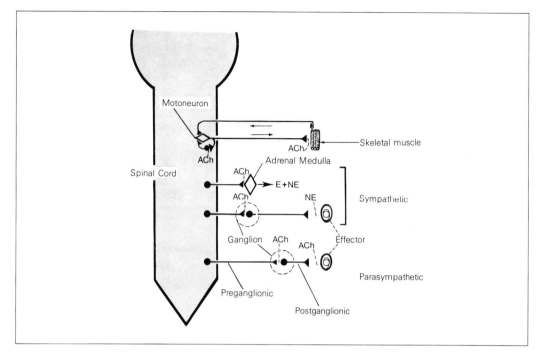

*Figure 4–17. Distribution of transmitters acetylcholine (**ACh**), norepinephrine (**NE**), and epinephrine (**E**).*

terminal, in which the norepinephrine can be recycled into storage vesicles and released again, or changed to an inactivate metabolite under the influence of the enzyme monoamine oxidase (MAO).

Adrenergic neurons or secretory cells are those that secrete norepinephrine (noradrenaline) as a transmitter; they include the following (figure 4–17):

1. Postganglionic fibers of the sympathetic system (with the exception of a few described under acetylcholine)
2. Secretory cells in the adrenal medulla, which are analogous to postganglionic neurons of the sympathetic system and secrete a mixture of norepinephrine and epinephrine
3. Some neurons in the CNS, particularly those arising in certain brainstem regions (locus ceruleus and reticular formation)

There is evidence linking the CNS effects of norepinephrine with such complex functions as alertness, learning, sleep, and mood.

Dopamine

Dopamine is the precursor of norepinephrine in neurons that secrete norepinephrine. Certain identified neurons, however, secrete dopamine directly. Important examples are some neurons in the basal ganglia whose degeneration is related to the movement disorder of Parkinson's disease. Dopamine-secreting neurons also project to the cerebral cortex and limbic system and may be involved in emotional responses.

Serotonin

Serotonin (5-hydroxytryptamine or 5-HT) may be an inhibitory transmitter secreted by neurons

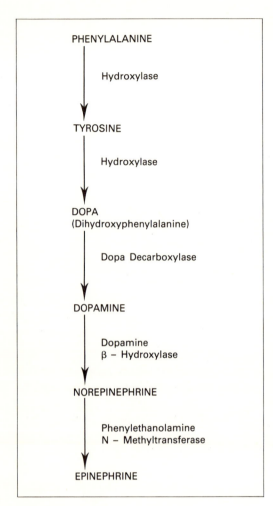

PHENYLALANINE

Hydroxylase

TYROSINE

Hydroxylase

DOPA
(Dihydroxyphenylalanine)

Dopa Decarboxylase

DOPAMINE

Dopamine
β – Hydroxylase

NOREPINEPHRINE

Phenylethanolamine
N – Methyltransferase

EPINEPHRINE

Figure 4–18. Synthesis of norepinephrine and epinephrine.

originating in particular regions of the brainstem (the midline or raphé nuclei). It has been tentatively linked with drugs that affect mood and cause hallucinations (e.g., LSD).

GABA and Glycine

Certain amino acids, in particular gamma-aminobutyric acid (GABA) and glycine, are found in high concentrations in the CNS and are effective in modifying neuronal discharge. While not yet conclusive, experimental evi-

dence suggests that they are CNS transmitters. GABA may be released as a transmitter by inhibitory interneurons in the brain (for example, in the cuneate nucleus of the somatic sensory pathway, in cerebellar cortex, and in cerebral cortex) and in the spinal cord (specifically, those interneurons responsible for presynaptic inhibition). Picrotoxin, a convulsion-inducing drug, may act by blocking the inhibitory effect of GABA. Glycine may mediate the effects of interneurons that cause IPSPs in spinal motoneurons. Strychnine, another convulsion-inducing drug, may act by antagonizing the inhibitory effects of glycine.

Drug Action at Synapses

Synaptic transmission can be modified with appropriate compounds that mimic or affect the action of native chemical transmitters. Such compounds can be either drugs or poisons (as we have seen for curare and related compounds, which block the acetylcholine receptor). There are several specific mechanisms for such drug effects.

1. *The compound can prevent release of the chemical transmitter from its presynaptic terminal.* Botulinum toxin, which sometimes appears in improperly prepared canned goods, blocks the release of acetylcholine, causing neuromuscular paralysis.
2. *The compound can block the postsynaptic receptor site by binding to it and competing with the transmitter.* Curare, tubocurarine, and alpha-bungarotoxin block the acetylcholine receptors in skeletal muscle. Another drug, atropine, blocks the acetylcholine receptors in smooth muscle and glands innervated by parasympathetic nerve fibers. This is the reason atropine eyedrops dilate the pupils: they block the pupillary constriction ordinarily caused by acetylcholine acting on the smooth muscle of the iris.
3. *The compound can block enzymatic breakdown of the chemical transmitter.* For example,

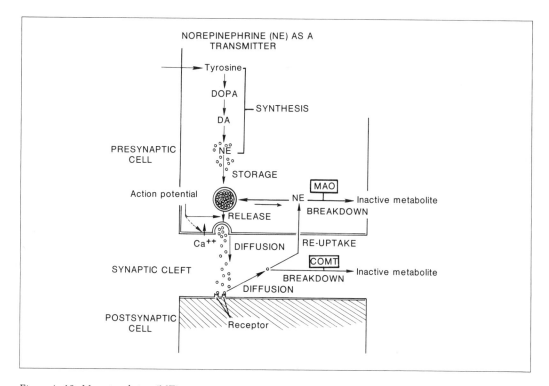

NOREPINEPHRINE (NE) AS A
TRANSMITTER

Figure 4–19. Norepinephrine (NE) as a neurotransmitter. **DA** *= dopamine.*

neostigmine blocks acetylcholinesterase and is, therefore, called an *anticholinesterase drug.* Since the acetylcholine no longer breaks down rapidly, it builds up to higher levels in the synapse. In the disease *myasthenia gravis,* weakness due to deficient cholinergic activity may be counteracted by administering such anticholinesterase drugs.

4. *The compound can block the re-uptake of the transmitter.* For example, the tricyclic antidepressant drugs, such as imipramine and amitriptyline, are thought to block the re-uptake of norepinephrine into presynaptic terminals. This allows more norepinephrine to accumulate in the synapse and counteracts depressive symptoms that, according to one theory, are due to norepinephrine deficiency.

The effect of a drug depends on which chemical transmitter system it affects and on the mechanism by which it works. There are, for example, drugs that block the receptor site for acetylcholine in striated muscle, others that block the receptor site for acetylcholine in smooth muscle, others that block the receptor site for norepinephrine in smooth muscle, and others that block the receptor site for dopamine in the brain. Obviously, these drugs can have radically different effects. In addition, both curare and neostigmine work on acetylcholine transmission on skeletal muscle, but since they work by different mechanisms, they have different effects: curare blocks the action of the transmitter, while neostigmine can enhance it.

Review Exercises

1. Diagram a neuromuscular junction, labeling the presynaptic terminal, synaptic vesi-

membrane. Diagram a neuron–neuron synapse similarly.

2. Describe the sequence of events during neuromuscular transmission.

3. Define end-plate potential (EPP) and miniature end-plate potential (MEPP) and describe their relationship.

4. Describe the changes in membrane potential and ionic conductance during neuromuscular transmission.

5. State the relationship between the EPP and the muscle action potential. Define threshold in this context.

6. Describe changes in membrane potential and ionic conductance associated with EPSPs and IPSPs.

7. Specify a value for synaptic delay, and describe the basis for it.

8. State the relationship of EPSPs and IPSPs to the initiation of an action potential in the postsynaptic neuron.

9. Describe the part of the motoneuron with the lowest threshold. State the approximate value of the threshold.

10. State in what ways postsynaptic potentials differ from action potentials.

11. Relate the characteristics of postsynaptic potentials to summation.

12. Compare postsynaptic inhibition with presynaptic inhibition.

13. State the transmitters at the following synapses: (1) neuromuscular junction (skeletal muscle), (2) autonomic ganglia, (3) sympathetic effectors, and (4) parasympathetic effectors.

14. Describe the effect of curare on neuromuscular transmission.

15. List four mechanisms by which drugs can act at synapses, and classify each as to whether it increases (potentiates) or decreases the postsynaptic action of the transmitter involved.

16. Explain why an anticholinesterase agent is sometimes given to counteract the residual effects of tubocurarine after surgery.

5. Somatic Sensation

How the nervous system converts touch, pressure, joint position, temperature, and painful stimuli into sensations

When a pacinian corpuscle receives an adequate stimulus, it sends nerve impulses along a peripheral nerve into the CNS. So do the various other receptors for touch, pressure, joint position, temperature, and pain that innervate the skin and other tissues of the body. Some of these sensory nerve impulses may have immediate reflex effects at a spinal-cord level. Others reach the brain where they elicit somatic sensations corresponding to the stimuli impinging on the body.

Receptor mechanisms were described in chapter 3. In brief, each receptor is most sensitive to a particular form of external energy (although some receptors associated with thin myelinated or unmyelinated nerve fibers respond to several forms of energy). This external energy is transduced (converted) into a localized and graded depolarization called a *receptor potential*. If the external stimulus is adequate in kind and intensity (an adequate stimulus), the receptor potential will trigger one or more action potentials. The action-potential frequency, along with the size of the receptor potential, tends to increase with greater stimulus intensity; both also decline with time during a maintained stimulus, a process called *adaptation*, which varies from rapid to slow.

Sensory-Nerve Conduction

After action potentials have been initiated in somatic sensory receptors, the next step is their conduction into the CNS via sensory (afferent) nerve fibers. Although the normal stimuli for these receptors are mechanical, thermal, and so on, the conduction process is more easily studied by stimulating the peripheral end of the nerve trunk with an electrical pulse, while recording the electrical response. When a peripheral nerve, such as the median or sciatic nerve, is stimulated electrically, numerous nerve fibers will respond with a nerve impulse. An electrode on or near the surface of the nerve can then record a compound action potential (so called because it is compounded from the ionic current of many individual action potentials). This technique has been used in the clinical testing of peripheral nerve function and in the physiological investigation of how different types of peripheral nerve fibers serve different functions.

Clinical Testing of Sensory-Nerve Conduction

Clinical testing may be indicated in order to determine whether the sensory nerve fibers are conducting at the normal velocity. An example of the procedure is shown in figure 5–1. A small electrical pulse applied to the finger stimulates sensory-nerve fibers, which travel centrally in the digital, and then the median nerve. A recording electrode on the skin of the wrist picks up the compound action potential of the median nerve at a point where the nerve comes close to the surface. The time interval, or *latency*, from stimulus to action potential in this example is 3 msec, while the distance between stimulating and recording electrodes is

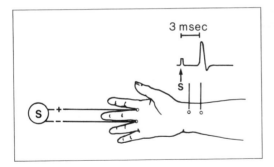

Figure 5–1. Testing conduction velocity in a sensory nerve. Stimulus applied over digital nerves elicits compound action potential, which can be recorded from skin over median nerve. S, stimulus. (Method described in Smorto, M.P., and Basmajian, J.V. [32].)

150 mm. The conduction velocity, then, is

velocity = distance/time = 150 mm/3 msec = 50 m/sec

The conduction velocity may be significantly

Figure 5–2. Recording the compound action potential of a peripheral nerve. Inset: Diagram of the experimental method. S, stimulating electrode; R, recording electrode (reference electrode is at right end).

reduced in certain conditions, for example, when the peripheral nerve is constricted by connective tissue (in the median nerve this is often associated with carpal tunnel syndrome).

Physiological Classification of Sensory-Nerve Fibers

Large peripheral nerves, such as the sciatic, have been removed from animals and stimulated at the peripheral end by an electrical pulse (figure 5–2). Near the other end, a recording electrode (a metal wire) picks up the electrical signs of the compound action potential. Its reference electrode is placed a few centimeters away, over a section of nerve that has been crushed; this ensures that no nerve impulses activate the reference electrode, so that it can provide an inactive baseline from which to measure activity under the recording electrode. Unlike an individual action potential in a single nerve fiber, the compound action potential increases in size as stimulus intensity is increased. This occurs because more and more nerve fibers are being activated, and the compound action potential represents a

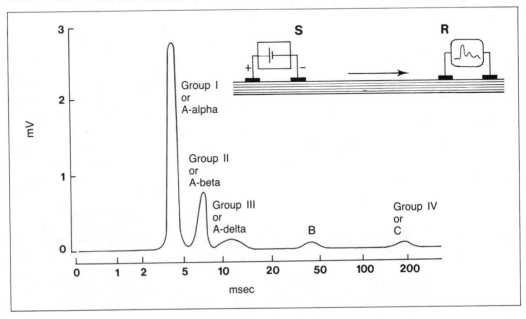

summation derived from these individual responses.

Also unlike the individual action potential, a compound action potential may have several peaks, because the nerve contains several groups of fibers with different conduction velocities. Imagine that the nerve fibers are runners in a race. As they cross the finish line (the recording electrode), the number arriving at any given fraction of a second after the starting gun are recorded on a chart. The next fastest group arrive bunched together and form a peak at the beginning of the chart. The next fastest group come in a few milliseconds later and form a second peak, and two or three more groups of stragglers provide the later peaks. The first and largest peak represents activity in the Group I fibers, the fastest conducting of all, with a velocity of around 100 m/sec. The next peak represents activity in the Group II fibers, not quite as rapidly conducting at about 50 m/sec. Peaks due to Group III (around 25 m/sec) and Group IV (about 1 m/sec) follow.

The peaks in a compound action potential have also been classified alphabetically. In general, Group I corresponds to the A-alpha peak; Group II to A-beta; Group III to A-delta; and Group IV to C. In addition to conduction velocity, the fiber groups differ structurally under a microscope: Group I are the largest-diameter myelinated fibers, Group II are large-diameter myelinated fibers, Group

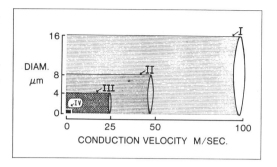

Figure 5–3. *Typical conduction velocity (in m/sec) and fiber diameter (in μm) for different groups of afferent nerve fibers.*

III are small-diameter myelinated fibers, and Group IV are unmyelinated fibers. The relationship of classification, conduction velocity, and fiber diameter is diagramed in figure 5–3.

Furthermore, and most important in studying somatic sensation, the different fiber groups have different functions. The Group I fibers carry signals from muscle and tendon receptors, which are important for controlling posture and movement; they will be discussed in chapter 9. The Group II fibers carry sensations from receptors for touch, pressure, vibration and joint position and movement, and are discussed below. The other two groups, which carry information about pain and temperature (as well as crude touch), will be discussed later in this chapter. Table 5–1 summarizes the types of afferent fiber and their major functions.

Table 5–1. *Afferent Fiber Types and Their Major Functions*[a]

Group	Type	Description	Typical Conduction Velocity and Range (in msec)	Major Functions
Group I	A-alpha	Largest-diameter myelinated	100 (72–120)	Muscle feedback from primary muscle stretch receptors and Golgi tendon organs
Group II	A-beta	Large-diameter myelinated	50 (36–72)	Touch, pressure, vibrations, and kinesthesis
Group III	A-delta	Small-diameter myelinated	25 (6–36)	Fast pain, temperature, and crude touch
Group IV	C	Unmyelinated	1 (0.5–2)	Slow pain, temperature, and crude touch

[a] *Source:* Adapted from W.D. Willis and R.G. Grossman. *Medical Neurobiology*, 2nd ed. St. Louis: The C.V. Mosby Co., 1977.

Discriminative Somatic Sensation and the Dorsal Column System

Discriminative somatic sensations are those that allow you to discriminate a nickel from a quarter by touch alone, for example. These sensations involve peripheral afferent fibers belonging to Group II and a CNS pathway including the dorsal columns of the spinal cord.

Tests of Discriminative Somatic Sensation

Several tests are described briefly below. All should be done with the subject's eyes closed. Similar tests are included in the neurological examination.

1. *Touch.* With a small, fine brush, lightly touch the skin over different areas of the body (fingers, palms, arms, back, and so on). Ask the subject to say when he feels the touch and where.
2. *Two-point tactile discrimination.* This is the ability to discriminate two closely spaced touch stimuli from a single one. Two points (pencil tips, calipers with tips covered, or a specially designed device called an esthesiometer) a few centimeters apart are lightly touched to the skin alternated with a single point. The subject says "one" or "two" as the points are felt. The distance between the two points is gradually reduced until the subject can no longer report accurately; this is the two-point tactile threshold. Compare this value on the fingers and on the upper arm or back. Notice that the finger threshold is close to the separation of the dots in Braille, which is 2.5 mm.
3. *Pressure.* Have the subject hold his palm upward and put different weights on his palm, one after the other. Have the subject say whether each weight is heavier or lighter than the previous one.
4. *Vibratory sense.* Strike a tuning fork so that it vibrates, and place its base on different parts of the body. Ask the subject to indicate when and where the vibration is felt.

5. *Position sense.* Move the subject's arm so that the angle at the shoulder joint is about 45 degrees from the vertical. Ask the subject to duplicate this angle with the other arm. Change the angle of the first arm, and ask the subject to duplicate the new angle.

Peripheral-Nerve Activity

Many of the skin and joint mechanoreceptors with specialized endings, such as the pacinian corpuscle, Meissner's corpuscle, Merkel's disc, hair-follicle receptor, and joint receptor, have large-myelinated Group II fibers. The responses of such afferent fibers have been described in chapter 3.

Stimulation of these peripheral nerve fibers by a mild electrical stimulus produces a sensation that appears to originate in the peripheral distribution of the nerve. Thus, a mild electrical shock to the median nerve at the wrist can give a tingle, tap, or slaplike feeling to the area of the hand innervated by the nerve.

The large-myelinated Group II afferents can be functionally separated from the small-myelinated and unmyelinated fibers in several ways.

1. They can be stimulated by external electrodes at a lower voltage.
2. They can be more rapidly blocked by a tourniquet, which reduces blood flow, causing ischemia. So discriminative touch may be blocked while pain sensation remains.
3. They are generally more slowly blocked by infiltration of local anesthetics, which can, therefore, block pain sensation while touch sensation remains.

Central Nervous System Activity

Group II fibers enter the spinal cord through the dorsal roots, and then ascend on the same side of the cord in the *dorsal columns*, which can be seen on the dorsal surface of the cord (figure 5–4). The dorsal columns are composed

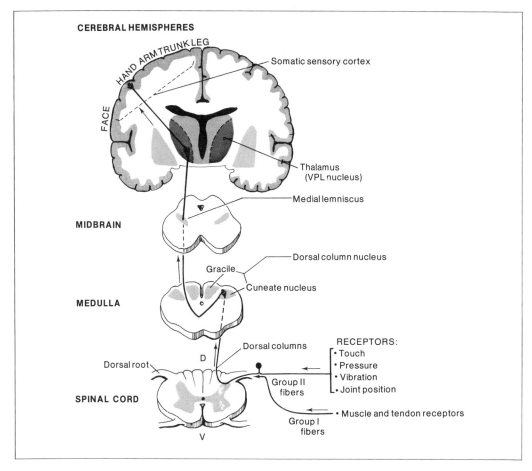

Figure 5–4. The dorsal column–medial lemniscal (DC–ML) pathway for somatic sensation. **D** = *dorsal;* **V** = *ventral.*

of the cuneate fasciculus (from the arms) and the gracile fasciculus (from the legs). They appear white because they are composed of myelinated nerve fibers. The fibers ascend and synapse onto neurons in the dorsal-column nuclei (the cuneate and gracile nuclei) in the base of the medulla just above the spinal cord. These neurons send fibers across the midline to the opposite side of the medulla, where they continue upward as the *medial lemniscus.* Thus, this system is often called the dorsal column system or the dorsal column–medial lemniscal (DC–ML) system.

Nerve fibers in the medial lemniscus synapse in a thalamic relay nucleus, the ventral posterolateral or VPL nucleus. Fibers from the VPL nucleus then project to the cerebral cortex, mainly to somatic sensory area I in the postcentral gyrus of the parietal lobe. Because of the crossing over in the medulla, the left half of the body is represented in the right postcentral gyrus, and the right half of the body is represented on the left postcentral gyrus (a contralateral representation).

Somatotopic Mapping. On the postcentral gyrus, the contralateral half of the body is represented by a somatotopic map, very much the way the world can be mapped on a sheet of paper. The body is mapped upside down, with the face below and the legs above. The cortical area for each body part reflects density of innervation and sensory function rather than skin area; thus, the hands and mouth (which have a high degree of discriminative touch sensation as shown by the two-point tactile test) have large representations, while the back has a small one. This mapping has been demonstrated physiologically by recording the electrical responses of the cortex to touch stimulation of different parts of the body. The electrical response, or evoked potential, to stimulation of the right hand is at a point about 7 cm to the left of the midline. The evoked potential to stimulation of the legs is best recorded from the part of the postcentral gyrus located on the medial surface of the hemisphere.

Receptive Fields and Surround Inhibition. The receptive field of a neuron in the somatosensory system is a particular area of skin within which an adequate stimulus causes the neuron to

Figure 5–5. The excitatory receptive field of an afferent fiber (a first-order neuron in the somatic sensory pathway).

respond. For example, a sensory nerve fiber may innervate a particular area of skin in the hand. An adequate stimulus within this area will cause excitation (increase in firing rate) of the nerve fiber. This area is called the *excitatory receptive field* of the nerve fiber (figure 5–5).

The size of the receptive field varies widely, from very small on the fingertips (a few square millimeters) to large on the back (several square centimeters). This size corresponds logically with the two-point tactile threshold on different parts of the body. If the receptive field of a fiber innervating the fingertips were large—2 cm square—it would cover the whole fingertip, and two points touching the fingertip would both activate the same nerve fiber, so it would be impossible to differentiate the two points from a single point applied anywhere on the fingertip. Since the two-point threshold for the fingertips is actually 2 to 3 mm, the diameter of the receptive fields on the fingertips must generally be less, so each point can excite a different fiber.

Fibers in the median nerve ascend in the cuneate fasciculus (part of the dorsal column) and then synapse onto second-order fibers in the cuneate nucleus. A microelectrode in the

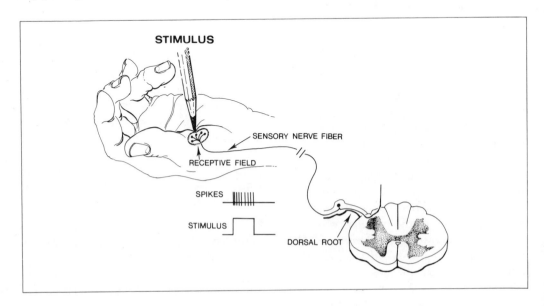

cuneate nucleus could record activity from these second-order fibers. They too have receptive fields in the arm and hand. Unlike the first-order sensory fibers, however, their excitatory receptive fields are surrounded by inhibitory areas, or inhibitory surrounds, stimulation of which decreases the firing rate of the nerve cells.

The mechanism of *surround inhibition* (also called *lateral inhibition*) is diagramed in figure 5–6. A table edge is stimulating neuron *b*, a first-order sensory fiber that ascends in the cuneate fasciculus, by pressing on its excitatory receptive field. Neuron *b* synapses directly with second-order neuron *b'* in the cuneate nucleus. Neuron *b'* is also activated by pressure from the table edge in its excitatory receptive field. Its neighbors, neurons *a'* and *c'*, however, have the table edge pressing their inhibitory surrounds, and respond with a decrease in firing rate (inhibition) below their ongoing spontaneous rate. These firing rates are graphed on the right of the diagram. Notice that the excited neuron (neuron *b'*) is surrounded on both sides

by inhibited neurons *a'* and *c'*. In reality, this diagram could be extended above and below the plane of the paper, and neuron *b'* would be ringed on all sides with inhibited neurons. This contrast between an activated neuron and its inhibited neighbors is considered important for precise localization of where the stimulus is and the ability to differentiate between two closely spaced stimuli, as in two-point tactile discrimination.

The diagram illustrates two synaptic mechanisms for surround inhibition:

1. *Postsynaptic inhibition.* A branch (axon collateral) of axon *b'* synapses with an inhibitory interneuron. Two such inhibitory interneurons then synapse onto neighboring neurons *a'* and *c'*, and inhibit them via postsynaptic inhibition (generation of an IPSP and reduction of firing rate).
2. *Presynaptic inhibition.* A branch of first-order fiber *b* synapses with an interneuron, which synapses with the axon terminal of neighboring fiber *c*. This axoaxonal synapse (from one axon terminal to another) is able to depress the release of transmitter from nerve fiber *c*, causing less activation of neuron *c'*.

Figure 5–6. A schematic representation of surround inhibition in the somatic sensory system.

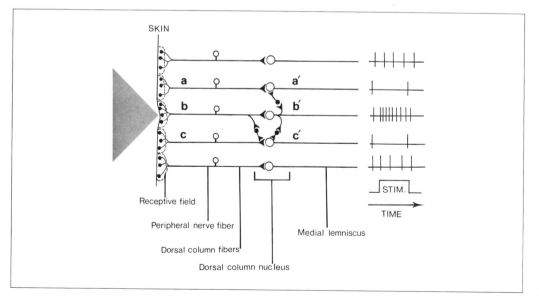

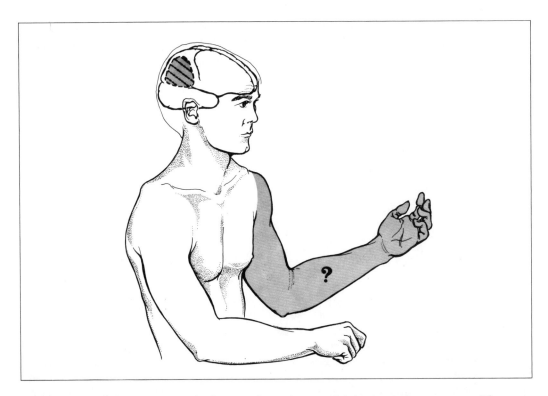

Figure 5–7. A possible consequence of a lesion to the right parietal association cortex.

In both cases, interneurons within the cuneate nucleus inhibit the neurons surrounding the most excited neuron and allow it to stand out.

The amount of surround inhibition can also be affected from higher centers in the brain. For example, nerve fibers have been traced from the somatic sensory cortex to the cuneate nucleus. Activation of the descending fibers can affect the inhibitory interneurons and alter the degree of surround inhibition in the cuneate nucleus. We can speculate that attention directed toward a fingertip might increase the amount of surround inhibition and increase the ability to differentiate between closely spaced stimuli. Similar mechanisms occur in the gracile nucleus, the VPL thalamic nucleus, and the postcentral gyrus.

Functions of Posterior Parietal Association Cortex. Posterior to the postcentral gyrus is a large area

of parietal lobe association cortex. This area has importance in the perception of bodily sensations. Anatomically, it receives direct or indirect projections from the postcentral gyrus. One indication of this function is given by the clinical observation that damage to the posterior parietal association cortex on the right hemisphere (figure 5–7) often leads to defective perception of sensations from the opposite (left) side of the body. The patient may fail to recognize the left side of his body as his own; he may fail to dress it, shave the left side of his face, or comb the left side of his hair. He may also fail to perceive objects to the left in his field of vision. A reasonable conclusion from this syndrome is that the posterior parietal association cortex normally functions to integrate somatic and visual sensation into a complete body image. Similar lesions on the left hemisphere may produce similar defects on the right side of the body, although often less marked and mixed with language disorders, since in most people the left hemisphere has dominant importance in language.

Pain Sensation and the Spinothalamic System

Pain and temperature sensation are carried from peripheral receptors (nociceptors for pain, thermoreceptors for temperature) by small-myelinated (Group III or Type A-delta) and unmyelinated (Group IV or Type C) fibers in the peripheral nerve. In the spinal cord, these fibers transmit activity to fibers that ascend in the lateral spinothalamic tract to several areas of the brain. For brevity we will refer here only to pain sensation, and we will call the whole CNS pathway the spinothalamic system.

Peripheral-Nerve Activity

Some of the peripheral stimuli that cause pain are:

1. Intense pressure on the skin, cutting, or other tissue damage. Tissue damage can release a polypeptide, such as bradykinin, that stimulates nociceptors.
2. Sustained muscle contraction that causes ischemia (lack of blood supply) to the muscle.
3. Other causes of ischemia to a muscle, for example, pain arising in the heart (angina pectoris) when a coronary artery is blocked.
4. Stretch of a visceral organ, for example, stretch of the temporal artery associated with vasodilation in migraine headache or distention of the ureter when it is blocked by a kidney stone.
5. An abnormal stimulus along the course of a peripheral nerve or its dorsal root causing pain that seems to originate in the peripheral region innervated by the nerve. An example is pain due to a damaged intervertebral disc (a so-called slipped disc) which exerts pressure on the dorsal root.

If you stub your toe, you will probably feel two phases of pain: an immediate sharp sensation, followed a short interval later by a more-generalized ache. These two phases are attributed to two sets of peripheral-nerve fibers conducting impulses at different speeds.

1. *Fast pain* is the more-immediate, sharp, and localized sensation. It is
   a. Carried by A-delta fibers, conducting at a velocity of about 25 m/sec and capable of activating the brain in a fraction of a second
   b. A sharp sensation, resembling a pinprick
   c. Localized to the area stimulated
   d. Brief
2. *Slow pain* is the delayed ache. It is
   a. Carried by C fibers, conducting at a velocity of about 2 m/sec and capable of activating the brain after a delay of a second or more
   b. An aching or burning sensation, more unpleasant than fast pain
   c. Diffuse, tending to spread from the immediate area stimulated
   d. Sustained, often outlasting the stimulus for some time

Central Nervous System Activity

The *lateral spinothalamic tract* and its origins and projections are schematically diagrammed in figure 5–8. A-delta and C fibers from nociceptors may go up or down a few segments after they enter the spinal cord and terminate in the dorsal horn of the cord. After one or more synapses, activity is relayed by dorsal-horn neurons which send axons across the anterior commissure to ascend in the lateral spinothalamic tract of the opposite side. These spinothalamic fibers terminate in several areas of the brain:

1. In the thalamus: the VPL nucleus, and a region posterior to it called the posterior nucleus (or nucleus ventralis caudalis)

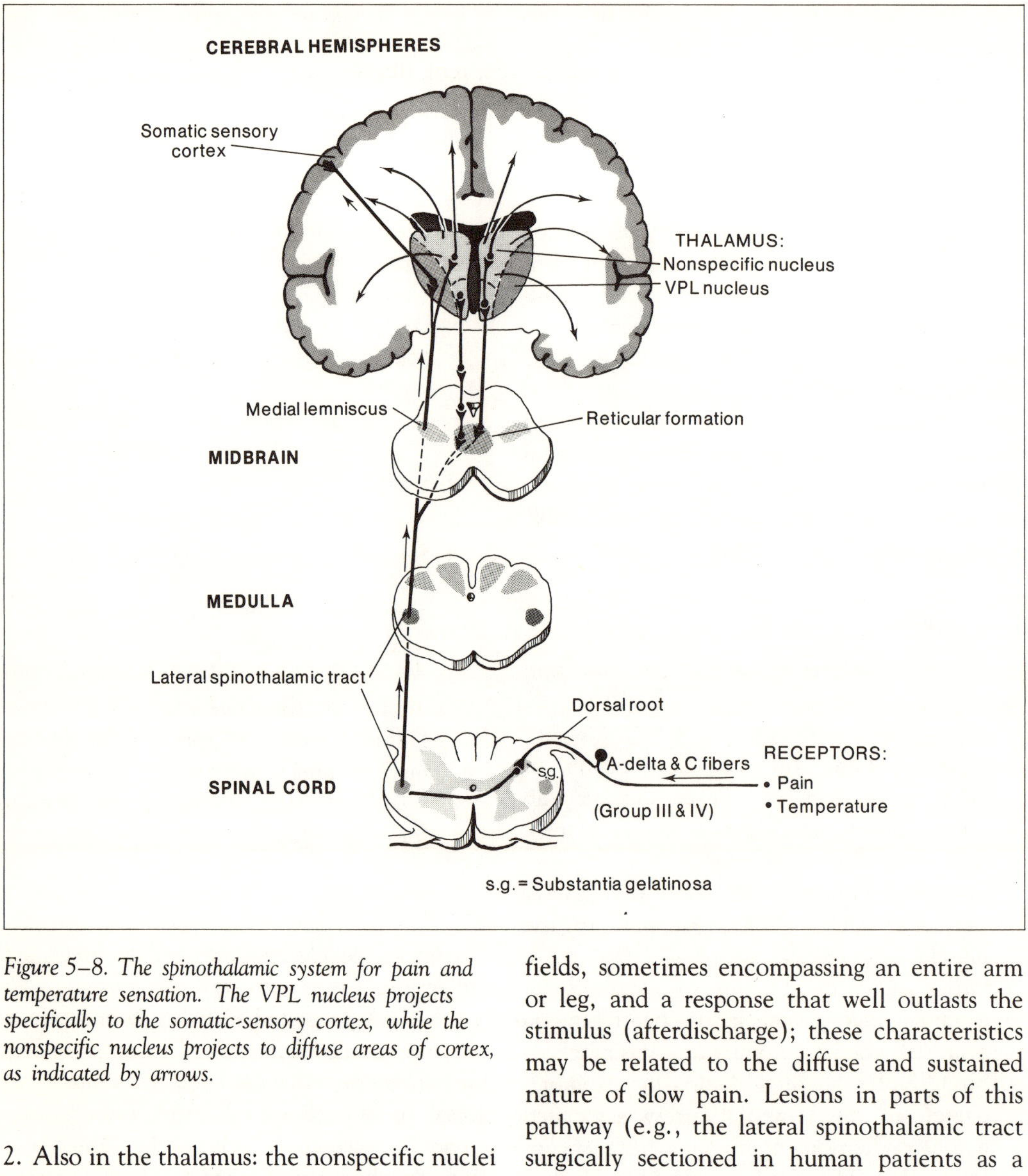

Figure 5–8. The spinothalamic system for pain and temperature sensation. The VPL nucleus projects specifically to the somatic-sensory cortex, while the nonspecific nucleus projects to diffuse areas of cortex, as indicated by arrows.

2. Also in the thalamus: the nonspecific nuclei (intralaminal and medial thalamic nuclei). Neurons in thalamic areas (1) and (2) project to the cerebral cortex.
3. In the brainstem: the reticular formation, a column of neurons in the central core of the brainstem.

Except for the VPL nucleus, many of the neurons in these areas have large receptive fields, sometimes encompassing an entire arm or leg, and a response that well outlasts the stimulus (afterdischarge); these characteristics may be related to the diffuse and sustained nature of slow pain. Lesions in parts of this pathway (e.g., the lateral spinothalamic tract surgically sectioned in human patients as a treatment for pain, and lesions in the reticular formation in animals) have been followed by a lack of responsiveness to pain stimuli. However, such lesions by no means always abolish pain. Furthermore, there are many complex pain phenomena that cannot be predicted by considering the anatomical pathway alone; these phenomena include referred pain and gating.

Referred pain is pain originating in one peripheral site that produces a sensation as if coming from another site, presumably due to the nature of the CNS projections. Pain signals from the viscera are often felt at superficial sites that share the same or nearby spinal segments. For example, cardiac pain is often referred to the chest and the inside of the left arm, areas supplied by the same or neighboring dorsal roots of the thoracic cord (the heart is supplied by segments T1 through T5). This phenomenon is often attributed to convergence of afferent fibers from the heart and from the skin onto the same neurons in the dorsal horn of the cord, which then project upward via the lateral spinothalamic tract (figure 5–9).

Gating is a mechanism by which pain signals within the spinal cord can be suppressed by activity in the large-myelinated Group II (A-beta) fibers due to touch, pressure, or electrical stimuli. For example:

1. Moving water and massage have been used to suppress pathological pain, as in the neurological disorder causalgia.
2. Pain arising from some peripheral site can sometimes be suppressed either by mild electrical stimulation of nearby skin or of a peripheral nerve supplying the area, both of

which are too mild to cause pain sensations themselves but can activate Group II afferent fibers.

This interaction could work by means of inhibitory interneurons, excited by the large-myelinated afferents in the dorsal horn of the cord, which then inhibit synaptic transmission (closing the gate) from the A-delta and C fibers to the dorsal-horn cells that give rise to the lateral spinothalamic tract.

There is also evidence that a similar inhibition in the dorsal horn could originate in either the somatic sensory cortex or the brainstem reticular formation, with descending activity carried by the reticulospinal and other tracts.

Pharmacological Relief of Pain

Drugs given for the relief of pain fall into several main categories.

Non-narcotic analgesics, such as aspirin and acetaminophen, may be useful in relatively minor painful conditions such as headache. There is evidence that they act peripherally to reduce the synthesis of prostaglandin, a substance that sensitizes free nerve endings to painful stimuli.

Local anesthetics, such as procaine, are used to block nerve conduction either in the peripheral nerve, the dorsal roots, or the spinal cord itself, depending on where the drug is applied. Consciousness is not normally affected.

Narcotic analgesics, such as morphine and codeine, are opiate drugs. They tend to subdue the affective reaction to the pain, even though the painful sensation may remain. They may also produce sedation, euphoria, and anxiety reduction. Opiate drugs have been studied by radioactive labeling of their antagonist naloxone. The labeled antagonist has been found to bind with opiate receptor sites in the dorsal horn of the cord, the reticular formation, and the nonspecific thalamic nuclei, and also in the hypothalamus and amygdala, areas associated with emotional and autonomic response pat-

Figure 5–9. Convergence as a possible mechanism for referred pain originating in the heart. Since the dorsal horn neuron shown usually signals path from the left arm, the same interpretation is given when it is activated by cardiac pain. **ST.Tr** = *(lateral) spinothalamic tract.*

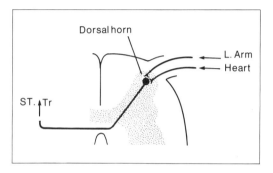

terns. The opiate receptors also act as binding sites for met-enkephalin, a five-amino-acid peptide found in pituitary tissue, and for endorphin, a polypeptide in human blood with the same five-amino-acid sequence. Enkephalin was found to have analgesic effects when injected into rat brain, suggesting that it may act as an endogenous analgesic substance.

Tranquilizing and antidepressant drugs may have favorable results in some patients by relieving the affective reaction to pain.

General anesthetics, such as enflurane, halothane, and nitrous oxide, may have a specific analgesic effect, and used during surgery at adequate levels, they also act to lower the level of consciousness. At appropriate doses these agents have been found to diminish the activity of neurons in the brainstem reticular formation and in the nonspecific thalamic nuclei [2], structures that function in the control of consciousness (chapter 11) as well as in the neural response to painful stimuli.

Psychophysiological Relief of Pain

Even in animals, the response to pain is not simply a matter of stimulus intensity alone; human contact and the presentation of food have diminished the response greatly under certain conditions. In humans, pain has been altered or relieved by several psychological and psychophysiological procedures.

Placebos are pills or other medications containing only presumably inactive ingredients, such as sugar. Susceptible individuals have reported less pain after taking a placebo medication—in one test, nearly half the number of patients who reported less pain after taking a narcotic analgesic. The power of suggestion induced by both the physician and the drug is probably behind the placebo effect.

Suggestion and hypnosis are also effective in selected individuals. Some patients have undergone dental work or even major surgery under hypnosis without anesthesia. Hypnosis is also sometimes used to alleviate the subjective experience of pain, including that of terminal illness. Such treatment requires special training and the recognition that individuals differ in hypnotic susceptibility.

Experimentally, a warning prior to an electric shock has been found to reduce the intensity of the response. Music or white noise (like that from an electric fan) coupled with the suggestion that this kind of sound reduces pain, has been found to reduce the pain response experimentally and during dental procedures, a phenomenon called *audio analgesia*. In clinical practice, relaxation training and anxiety reduction have been found to reduce pain intensity in some instances. Both methods are incorporated in various systems of natural childbirth education; factual information about childbirth, relaxation training, and breathing exercises cause a decrease in anxiety and an increase in the feeling of control.

Review Exercises

1. Describe methods for (1) clinical and (2) experimental measurement of peripheral-nerve-conduction velocity.
2. Classify the sensory fibers in peripheral nerve according to their group, alphabetical type, conduction velocity, and functions.
3. List and briefly describe the functions of the DC-ML system for discriminative somatic sensation.
4. Define the excitatory receptive field for a neuron in the somatic sensory system.
5. Define surround inhibition for such a neuron. Describe the synaptic mechanisms involved.
6. Define and describe the somatotopic mapping on the postcentral gyrus.
7. Describe the function of the posterior parietal association cortex.
8. List the structural and functional differ-

ences between the DC-ML and the lateral spinothalamic system.

9. List the differences between fast and slow pain.

10. Define referred pain. Describe a possible mechanism for it.

11. Define gating. Describe a possible mechanism for it.

12. List and briefly describe pharmacological and psychophysiological means of pain suppression.

6. Auditory and Vestibular Function

How neural mechanisms function in hearing and balance

Sound Waves and the Ear

Sound Waves

We all are familiar with wave motion in water: if you move your hand back and forth underwater, it generates waves that radiate outward from the point of origin. A loudspeaker cone in an audio system generates waves in air in a similar way (figure 6–1). The air molecules in front of the cone are compressed by each forward motion of the cone and decompressed (rarified) by each backward motion. If the speaker cone is moving back and forth in a simple sine wave pattern, alternating regions of high and low air pressure will radiate outward, and at any one instant, the air pressure will be a periodic function of distance as shown in figure 6–1. These pressure variations are called *sound waves* and are propagated at a velocity of 344 m/sec.

A microphone placed at some fixed distance from the loudspeaker will pick up the variations of sound pressure and convert them to electronic signals, which can be displayed as shown in figure 6–2. The simplest sound corresponds to a sine wave and is heard as a pure tone; although rare in nature, this can be produced by a soft whistle, a tuning fork, or an electronic oscillator. A pure tone can be characterized by its amplitude and by its frequency. A greater amplitude of movement by the loudspeaker produces a sound of greater amplitude or intensity, which is heard as a louder tone. The intensity is commonly measured by first cal-

culating the ratio of the pressure oscillations, P (in dynes/cm^2), to a reference value, P_0 (which was chosen to be near hearing threshold for a young person with good hearing). Then the logarithm of that ratio is multiplied by 20, to yield the sound pressure level, L, in units called decibels (dB):

$$L \text{ (dB)} = 20 \log \frac{P}{P_0}$$

The frequency, F, of a pure tone is the number of cycles per second (expressed as c/s, cps) or hertz (Hz). Sound frequency corresponds to what we hear as musical pitch. A tone of frequency 256 Hz will be heard as middle C. Increased frequency corresponds to a higher-pitched sound. Doubling the frequency from 256 Hz to 512 Hz will generate a tone one octave above middle C. Another way of characterizing the sound is by measuring the time between recurring peaks—the period T (where $T = 1/F$).

Musical instruments and the human voice often produce complex periodic sounds, which are periodic (they repeat themselves in a regular period of time) but more complex than sine waves. An example is shown in figure 6–3, where a tone of frequency F is combined with another tone of frequency $2F$ to produce the complex periodic sound in line C. The pattern in line C repeats itself periodically at the lower frequency F. Such sounds can be produced by combinations of frequencies F, $2F$, $3F$, $4F$, and

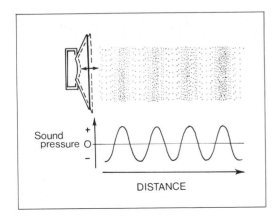

Figure 6–1. Sound waves in front of an oscillating loudspeaker cone, at one instant in time.

so on; the higher frequencies are called harmonics or overtones of the fundamental frequency *F*. These harmonics lend a quality (or timbre) that makes voices and musical instruments unique (a middle C on a piano sounds quite different from a middle C on a cello; both

*Figure 6–2. A microphone (center) converts sound pressure waves into electrical voltage, which can be displayed as it changes in time (lower right). The sound shown is a pure tone (sine wave). **T** = the period of the sine wave (the duration in seconds of each cycle). The vertical distance between peaks is a measure of amplitude.*

can have the same fundamental, but with a different combination of harmonics).

In addition to pure tones and complex periodic sounds, another type of sound is called *noise*. Noise has no recurring periodicity and is composed of random combinations of numerous frequencies. Vacuum cleaners produce good examples of noise.

Sound Conduction Through the Middle Ear

The middle ear (figure 6–4) is an air-filled space between the eardrum or tympanic membrane and the inner ear or cochlea. The three small middle-ear bones or ossicles (malleus, incus, and stapes) connect the tympanic membrane (eardrum) with the oval window (a membrane in the cochlea). The middle ear has three major functions: amplification, protection, and equalization. These functions may be expressed by the mnemonic APE.

Amplification. The middle ear amplifies sound-induced vibrations so that when they reach the oval window they are strong enough to overcome the impedance of the fluids that fill the cochlea. These fluids impede (or resist) vibrations due to sound, just as a moving paddle meets more resistance in water than in air. Fortunately, the amplification of sound vibra-

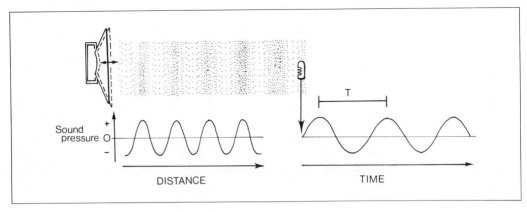

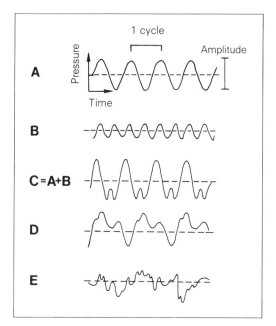

Figure 6–3. Sound pressure plotted against time. (A) A pure tone, of frequency F c/s or Hz. (B) A pure tone, of frequency 2F Hz. (C) A complex periodic sound, which can be derived from the combination of A and B. (D) A complex periodic sound from a musical instrument. (E) Noise.

tions that takes place across the middle ear is enough to overcome this impedance. The vibrations are transmitted by the mechanical linkage from the tympanic membrane through the ossicles to the footplate of the stapes in the oval window. Most of the amplification is due to the decrease in area between the tympanic membrane and the footplate of the stapes. The effective area of the stapes footplate has been estimated as only 1/14 the area of the tympanic membrane (figure 6–5). Since pressure equals force per unit area, this results in an amplification (by about 14 times) of the pressure. A small additional contribution may be made by the lever arrangement of the ossicles.

Protection. The middle-ear muscles protect the cochlea against sustained loud sounds. Loud

sounds elicit reflex contractions in the two middle-ear muscles, tensor tympani (inserted on the malleus) and stapedius (inserted on the stapes). The combined contraction increases the rigidity of the ossicular chain and reduces transmission, mainly for lower-frequency sounds (below about 1,000 Hz). Since the latency of the middle ear reflex is about 50 to 150 msec, there is no protection against very brief, sudden sounds, but the cochlea is provided some protection against sustained loud sounds, which might be attenuated by as much as 40 dB.

Equalization. The eustachian tube connects the middle ear to the pharynx and is opened during swallowing, yawning, and sneezing. At those times, it provides an open passageway between the middle ear and the atmosphere which allows steady pressure levels on both sides of the eardrum to equalize. Such equalization is necessary for the most efficient transmission of sound. During air travel, if cabin pressure becomes appreciably higher or lower than mddle-ear pressure and the eustachian tube is not opened, sound transmission across the middle ear is reduced, and pain may result as well.

Bone Conduction

In addition to the middle-ear pathway, sound can also be conducted to the cochlea directly through the bones of the skull, a process called *bone conduction.* When you hear yourself speak or hum, it is through a combination of air conduction (via the air, tympanum, and middle ear) and bone conduction (directly from the oral cavity through the skull to the cochlea). The reason your voice sounds different on a tape recorder is that only air conduction is involved. Bone conduction can be tested by placing a vibrating tuning fork or other object on the head.

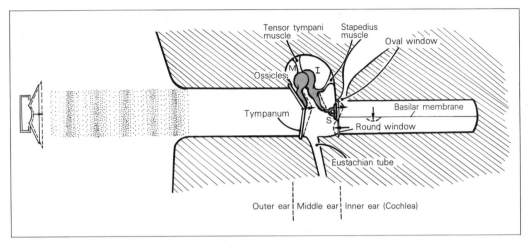

Figure 6–4. Outer, middle, and inner ear, showing the effect of sound waves at a particular instant in time. M = malleus (hammer), I = incus (anvil), S = stapes (stirrup). A compression wave momentarily causes an inward movement (shown by dashed lines) of the tympanic membrane, transmitted via the three middle-ear ossicles to oval window, basilar membrane, and round window. Cochlea is viewed as if unrolled to form a tube 3.5 cm long, then sectioned longitudinally and simplified as described in text.

Cochlear Mechanics

Structure of the Cochlea. Although the cochlea resembles a spiraling snail's shell, for simplicity it can be viewed unrolled to form a tube about 3.5 cm long (figure 6–4). The tube is filled with fluid (perilymph and endolymph) and divided lengthwise into two canals by the basilar membrane. The middle ear is separated from the upper canal by the oval window and from the lower canal by the round window (both windows are membranes). At the far end, both canals are connected by a small opening called the helicotrema. The basilar membrane supports the organ of Corti, a sensory structure containing auditory receptors (hair cells) and auditory nerve endings.

Traveling Waves in the Cochlea. When a tone causes the stapes to push inward on the oval window, the momentary increase in pressure in the upper canal causes a downward movement of part of the basilar membrane. The extra pressure transmitted to the lower canal is then relieved by a bulging of the round window into the middle ear (figure 6–4). After its downward displacement, the basilar membrane then rebounds upward; during a sustained tone, the basilar membrane moves up and down at the same frequency as that of the tone.

Stroboscopic photographs show that the up-and-down movement of the basilar membrane is transmitted as a traveling wave toward the end of the cochlear tube (figure 6–6). (It may help to visualize a long elastic band fastened to the wall at its far end, while your hand vibrates the near end up and down and generates waves that travel toward the far end). The up-and-down movement (vertical displacement) reaches a maximum amplitude at a particular distance along the basilar membrane and then dies away. The distance depends on the frequency of the stimulus tone:

Frequency	Distance at Maximum Amplitude
high	short (near oval window)
moderate	medium
low	long (near helicotrema)

A low tone involves a longer stretch of basilar membrane than a high tone. The envelopes

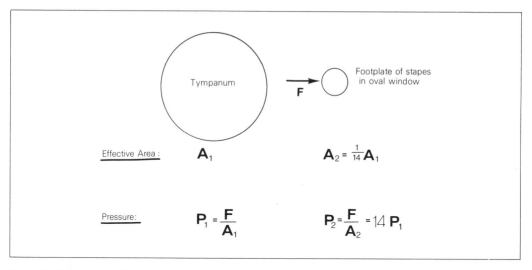

Figure 6–5. Schematic illustration of area difference between tympanum and footplate of stapes, and the accompanying pressure amplification between the two structures. The area ratio shown (14) illustrates the principle of pressure amplification, but varying estimates of this ratio have been given.

(outer bounds) of traveling waves for some frequencies are shown in figure 6–7. This mechanical behavior of the basilar membrane is largely due to its structure—narrow and stiff near the oval window, becoming wider and more flexible toward the helicotrema. A tone of a particular frequency (pitch) will cause maximal displacement at a particular place along the basilar membrane. This is one aspect of the *place theory* of hearing.

Neurophysiology of Audition

Transduction in Auditory Receptors

The hair cells, located on the basilar membrane and within the organ of Corti, are the auditory receptors. A tonal stimulus causes maximal up-and-down motion at a particular location along the basilar membrane. As seen in cross section (figure 6–8), both the hair cells that rest on the basilar membrane and the stiff tectorial mem-

brane that contacts the tips of the sensory hairs participate in this motion. The basilar membrane and the tectorial membrane act as if independently hinged at one wall of the coch-

Figure 6–6. Traveling waves along the basilar membrane shown at (A) one instant of time, in a three-dimensional representation, and (B) at four successive instants of time, from left to right, with basilar membrane represented only as a line corresponding to the heavy midline in A. Dashed lines show the limits of displacement, or the envelope, of these waves all along the basilar membrane, and arrows point to the maximal value of the envelope for the particular stimulus frequency used.

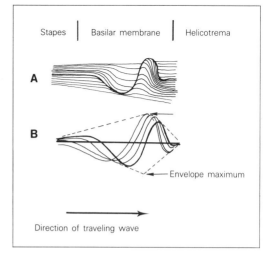

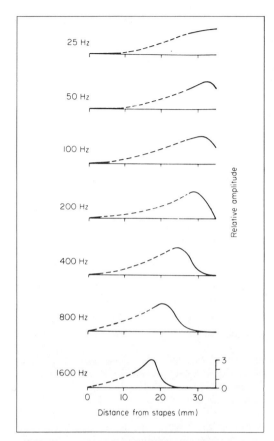

25 Hz

50 Hz

100 Hz

200 Hz

400 Hz

800 Hz

1600 Hz

Relative amplitude

0   10   20   30
Distance from stapes (mm)

Figure 6–7. Displacement envelopes for the stimulus frequencies shown beside each curve; only the positive half of the envelopes are shown. From Bekesy, G. Experiments in Hearing. New York: McGraw-Hill Book Co., 1960. Reprinted with permission.

lear tube; their vertical motion is converted into a horizontal shearing force exerted on the hairs of the hair cells. The hair cells behave like specialized mechanoreceptors, transducing the horizontal force into a receptor potential.

The receptor potential from numerous hair cells is the probable origin of an electrical wave, called the *cochlear microphonic* (CM), which can be recorded in or near the cochlea. With a pure tone (sine wave) stimulus, the cochlear microphone is a sine wave of identical frequency, similar to the electrical output of a microphone exposed to the same sound.

Neuronal Responses in the Auditory System

Anatomy of the Auditory Pathway. Auditory nerve fibers have their cell bodies in the spiral ganglion, and their peripheral endings synapse on the hair cells (figure 6–8). Their central endings synapse in the medulla at the cochlear nuclei. The auditory pathway ascends through several nuclei in the brainstem and thalamus to the cerebral cortex (figure 6–9).

Auditory Nerve Fiber Responses. Auditory nerve fibers end at synaptic structures at the base of the hair cells. Receptor potentials in the hair cells somehow elicit action potentials in the nerve fibers. A stimulus of a particular frequency will cause maximal movement at a particular place along the basilar membrane, which will activate a group of auditory nerve fibers that innervate that area of basilar membrane. Thus another aspect of the place theory of hearing is that *a given sound frequency will be represented in the CNS by a particular group of activated nerve fibers.*

The electrical response of individual auditory nerve fibers can be studied using tonal stimuli and microelectrode recording techniques. For example, as shown in figure 6–10, a fiber might respond to low-intensity tones of 1,600 Hz, while low-intensity tones above and below that frequency have no effect. The *best frequency* of this fiber is 1,600 Hz, and the frequency corresponds to the place innervated by the fiber on the basilar membrane. The fiber can also be excited by adjacent frequencies of greater intensity, particularly on the low-frequency side. A plot of all the frequency–intensity combinations that yield above-threshold responses is called the *tuning curve* of the fiber; it resembles the receptive field of a receptor on the skin or the retina. The tuning curves of two representative fibers are shown in figure 6–10; if a larger sample of tuning curves were shown, they would cover the entire range of audible frequencies. Increased sound intensity, corres-

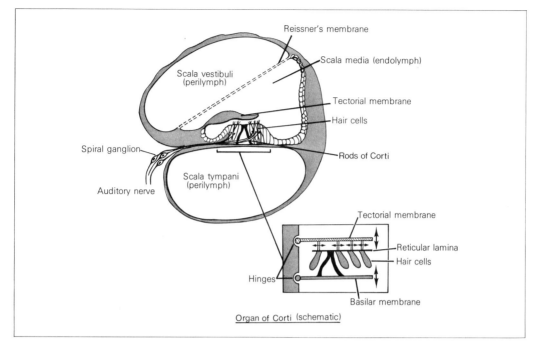

Figure 6–8. Cross-section of the cochlea. Inset shows a schematic diagram of the organ of Corti, and illustrates a possible mechanism for stimulation of the hair cells.

ponding to increased loudness, causes increased spike frequency for sounds within the neuron's tuning curve, as shown in figure 6–10 (bottom, right). Such an increase also adds or recruits new neurons to those that are responding.

Excitation of auditory nerve fibers is associated with upward motion of the basilar membrane. For a steady tone of low frequency (below about 2,000 Hz), nerve impulses tend to occur at a particular phase of the stimulus sine wave (figure 6–11). Such behavior is called *phase locking.* Although each fiber may not fire on each cycle, a group of neighboring fibers will tend to reproduce the periodicity of the stimulus in the timing of their impulses. There is evidence that such timing contributes to pitch perception for lower frequency sounds. The phase-locked response of groups of nerve fibers and its contribution to pitch perception is described in the volley theory of hearing.

Sound Localization. One can easily confirm that a person whose eyes are closed can localize the origin of a finger-snap, key-ring jingle, or other sound. The automatic tendency to attend to a sound source has obvious survival value in a world where either friends or enemies may be hidden in the forest, or where automobiles may appear from either end of a street you are about to cross. There are two major mechanisms involved; both are illustrated in figure 6–12 for a brief "tap" or "click" sound:

1. The sound arrives first at the ear nearest the source. The *time delay* between the two ears is estimated by dividing the difference in path length (in meters) by the speed of sound (in m/sec).
2. Because the farther ear is partially blocked by the head, there is an *intensity difference* between the two ears, the sound being less intense at the farther ear.

For the neural basis of localization, we focus on the superior olivary nucleus in the medulla, the first place where nerve fibers from left and

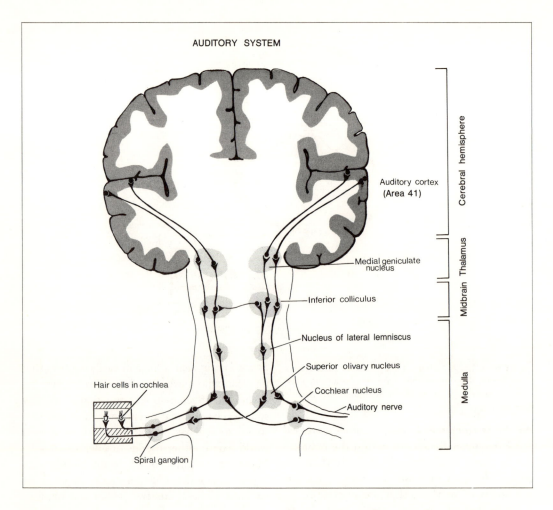

Figure 6–9. The auditory pathway (schematic). Only representative connections are shown.

right cochlear nuclei converge on the same cells. These cells receive a binaural input (from both ears), and respond precisely to particular combinations of time or intensity difference. One neuron, for example, might respond best to time delays of 0.4 msec (left ear preceding right), corresponding to a sound source a few degrees of angle to the left of the midline. Another neuron might respond best to a sound source 90 degrees to the right of the midline, and so on. Various left-right intensity differ-

ences also are preferred by different neurons. Thus sound localization can be coded in terms of which group of neurons in the superior olivary nucleus becomes most active. A similar coding occurs in cells of the inferior colliculus.

For low-frequency tones, the time delay effect involves phase-locked nerve impulses from left and right cochlear nuclei, similar to those shown in figure 6–11 for the auditory nerve. Phase-locked nerve impulses from the ear nearest the sound source slightly precede those from the other ear, and these time delays result in the firing of a particular group of cells which receive a binaural input.

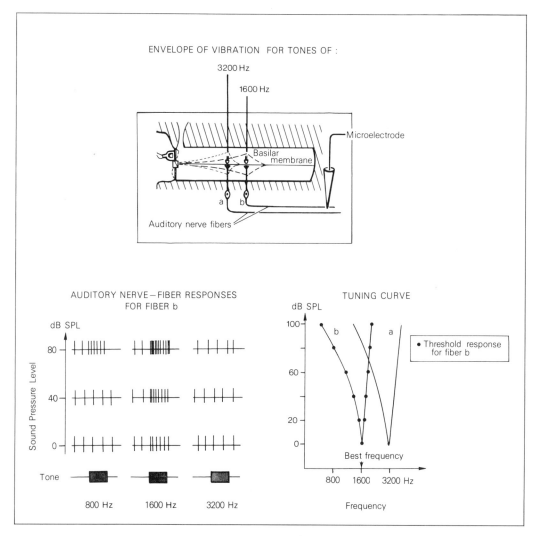

*Figure 6–10. Auditory-nerve-fiber responses (spikes) to tones of various frequencies and intensities. Left: Fiber **b** responds to a very low-intensity tone (near 0 dB SPL) of 1,600 Hz with a small increase in spike rate (a threshold response). An 800 Hz tone requires much greater intensity (80 dB) to elicit such a response. Right: Frequency-intensity combinations giving threshold responses for fiber **b** and fiber **a**. Each fiber has a unique best frequency, fiber **b** at 1600 Hz (arrow). Top: Traveling wave envelopes for 1,600 and 3,200 Hz tones, showing that auditory nerve fibers **a** and **b** innervate different areas of basilar membrane. Microelectrode can be used experimentally to record spikes.*

Auditory Cortex. At the upper end of the auditory pathway, nerve impulses are relayed from the medial geniculate nucleus (in the thalamus) to the primary auditory cortex (on top of the temporal lobes). The primary auditory cortex is also called Heschl's gyrus and Brodmann's area 41. It is connected to surrounding auditory association areas (Brodmann's areas 42 and 22).

Neuronal response characteristics are probably more complex in the auditory cortex than at lower levels of the pathway. Some cortical

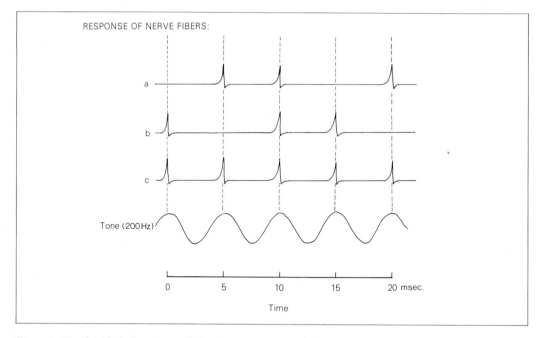

Figure 6–11. Phase-locked response of two representative auditory nerve fibers, **a** *and* **b,** *to a 200 Hz tone. In this example, each fiber tends to fire at the same phase of the sinusoidal stimulus—the peak, as shown by the dashed lines. A summation of the activity of fibers* **a** *and* **b** *would provide a signal (**c**) having the same period—5 msec—as the stimulus.*

neurons do not give a sustained response to a steady, pure tone [11]. Instead, they respond more reliably to changes in sound amplitude, such as the onset or offset of a sound, or to changes in sound frequency (a frequency-modulated sound). Such changing stimuli are more similar to the complex sounds that make up animal communication and human speech.

Because of extensive crossing fibers in the auditory pathway, Brodmann's area 41 receives binaural input (from both ears), although there is some predominance of input from the contralateral ear. Thus destruction of this area on one side does not cause complete hearing loss but only partial hearing loss in the contralateral ear. Deafness would be caused by bilateral destruction of Brodmann's area 41, but this condition is rare.

Hearing Sensitivity and Hearing Loss

Normal Hearing Sensitivity. When the frequency range of hearing is measured with pure tone stimuli, it ranges from a low of about 20 Hz to an upper frequency of 20,000 Hz (figure 6–13). This range is for a child with normal hearing; with increased age, the upper-frequency limit is reduced (to about 10,000 Hz at age 50). The *threshold of hearing* (auditory threshold) is the lowest sound pressure at which a tone is just audible; it varies with the frequency of the tone, reaching a minimum (hearing is most sensitive) between 1,000 and 4,000 Hz. At 1,000 Hz, the ideal threshold of hearing is a sound pressure (measured between upper and lower peaks of the sound wave) of 0.0002 dyne/cm^2, which has been established as the reference pressure (P_0) for measurements of sound intensity, L, in decibels (dB). For other sound pressures, P, the intensity can then be expressed as:

$$L \text{ (dB)} = 20 \log \frac{P}{P_0}, \text{ where } P_0 = 0.0002 \text{ dyne/cm}^2$$

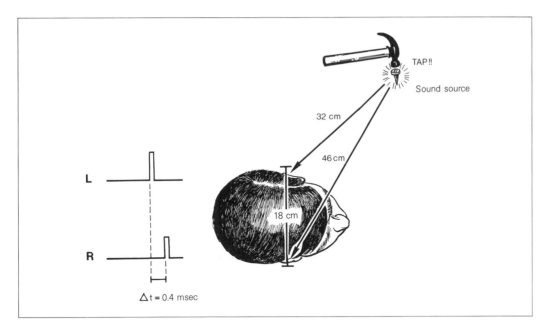

Figure 6–12. Localization of a transient sound. Path lengths of the sound are shown in centimeters. Sound-pressure waves as they arrive at both ears are shown in simplified form: note time delay Δt and amplitude difference between left (L) and right (R) ears.

Decibels calculated on this basis are called *decibels sound-pressure level* (dB SPL). In the most-sensitive hearing range, if the threshold of hearing is at a sound pressure, P, equal to P_0, then the threshold sound intensity is expressed as:

$$20 \log \frac{P}{P_0} = 20 \log 1 = 0 \text{ dB SPL}$$

For frequencies on either side of the 1,000 to 4,000 Hz range, greater sound intensities are required to reach the threshold of hearing.

The range of sound intensities present in everyday life is immense. Although under ideal conditions, a tone of zero dB SPL can just be heard, normal conversation ranges from 40 to 60 dB SPL, heavy traffic on a city street or amplified music on a dance-floor is approxi-mately 70 to 80 dB SPL, and a pneumatic drill is about 120 dB SPL. Auditory discomfort (tickle, touch, or pain sensations in the ear) is produced by intensities of 120 to 140 dB SPL.

Within the frequency range of greatest sensitivity (1,000 to 4,000 Hz), we can discriminate between tones only 2 to 6 Hz apart. The human speaking voice overlaps this frequency range, extending from 200 Hz (slightly below middle C) to 4,000 Hz.

Audiometry: The Measurement of Hearing. An *audiometer* is a device used for testing hearing by providing tones at various frequencies to each ear separately, generally via earphones. At each frequency, intensity is varied until the threshold of hearing is found. Results are plotted in the form of an *audiogram*, a graph of hearing loss as a function of frequency (figure 6–14), in which zero dB is commonly defined as the average threshold intensity for each frequency, obtained from a group of normal subjects. Hearing loss is then defined as the number of decibels above this value required to reach the threshold of hearing for the particular

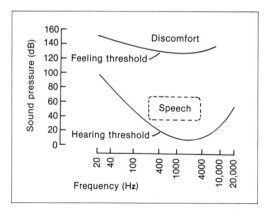

Figure 6–13. Hearing sensitivity. Frequency and intensity range of ordinary speech indicated by the dashed outline.

ear being tested. These are relative units, which should not be confused with dB SPL.

Hearing may be tested for two types of sound conduction. *Air conduction* refers to sound

Figure 6–14. An audiogram of a patient with hearing impairment, for stimulation of one ear. Bone conduction is normal (near the 0 dB hearing level), but air conduction is not, suggesting some degree of conduction deafness.

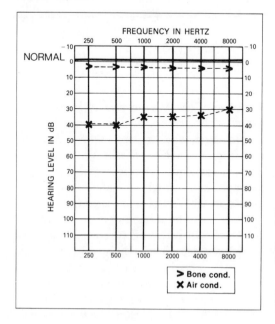

counducted through the air, transmitted through the tympanic membrane and the middle-ear ossicles to the cochlea. *Bone conduction* refers to sound conducted from a vibrating source placed on the mastoid or other bone, conducted directly through the skull to the cochlea. Although bone conduction is less efficient, the pattern of vibration set up in the cochlea is the same as that for air conduction.

Hearing loss falls into two major categories. (1) *Conduction* or *middle-ear deafness* is indicated when air conduction is impaired but bone conduction is normal. This could be due, for example, to calcium deposits making ossicles rigid, a condition called otosclerosis, which can often be treated surgically. The normal bone conduction indicates that the cochlear receptor mechanisms and the neural pathway is functional. (2) *Sensorineural deafness* is indicated when both air conduction and bone conduction are impaired and involves impaired function either in the cochlea or in the auditory pathway. For example, certain antibiotics may cause degeneration of cochlear hair cells; continued exposure to an intense high-pitched sound could lead to degeneration of hair cells in the corresponding part of the basilar membrane (near the oval window) and hearing loss specifically for the frequencies in that sound; and neural lesions in the brainstem can interrupt the auditory pathway.

Brainstem Auditory Evoked Response. Electrical responses can be recorded from the brainstem auditory pathway with electrodes on the scalp. The stimulus is generally a click presented to left or right ear via earphones. The electrodes are metal disks, like those used in recording the EEG, attached with conducting paste on the skin at the top of the head (vertex) and the ear. The voltage between the two electrodes is greatly amplified, and the response to 1,000 or more clicks is averaged by computer electronics. With this technique, we can record a complex electrical potential originating in the brainstem auditory pathway (figure 6–15). This

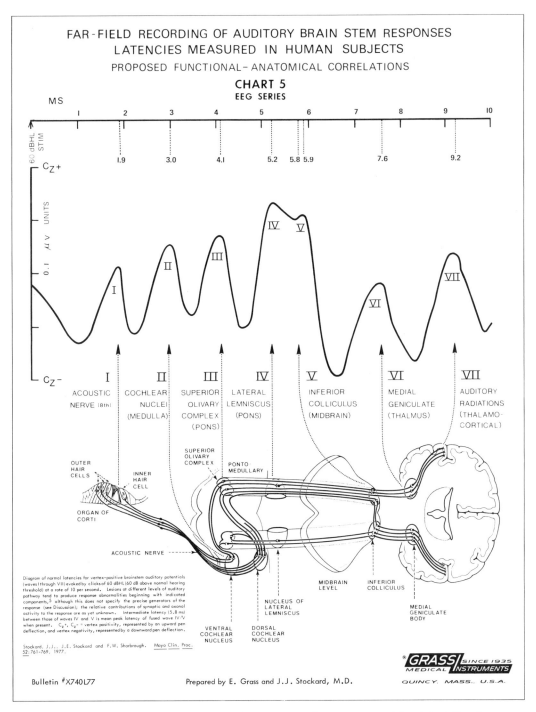

Figure 6–15. A normal brainstem auditory evoked response to clicks of 60 dB above normal hearing threshold presented at the onset of the trace. Total length of trace is 10 msec; normal latencies of numbered peaks are shown at dotted lines. An upward movement of the trace represents positive potential at the vertex electrode, as averaged over several thousand presentations of the click. Although precise neuronal origin of each wave is uncertain, lesions at different levels of auditory pathway diagrammed below tend to produce response abnormalities beginning with indicated waves. Courtesy of E. Grass and J.J. Stockard, and Grass Medical Instruments, Quincy, Mass. Reprinted with permission.

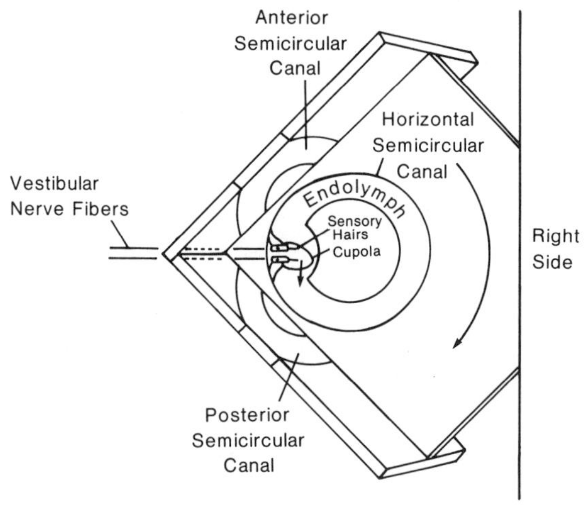

Figure 6–16. The orientation of the right semicircular canals as seen from above, and the response of the horizontal canal to rotations. As head rotates to right (clockwise), there is relative movement of fluid (endolymph) to left (counterclockwise).

potential, called the brainstem auditory evoked response or simply the brainstem evoked response, is composed of a sequence of low-amplitude (less than one microvolt) potentials during the initial 10 msec following click stimulation. These waves have been found to be altered in various neurological disorders affecting the brainstem auditory pathway. On this basis, wave I is thought to represent activity in the auditory nerve and wave V activity in the midbrain portion of the auditory system. The brainstem auditory-evoked response has been found useful in evaluating hearing loss and suspected lesions of the brainstem. It also demonstrates the rapid sequence of events in normal hearing: transmission from earphones to inferior colliculus takes only about 6 msec.

The Sense of Balance

The sense of balance is closely related to the sense of hearing: its sensory organ, the vestibu-lar apparatus, is located next to the cochlea in the temporal bone, and its nerve, the vestibular nerve, runs adjacent to the auditory nerve (together they make up cranial nerve VIII).

The Semicircle Canals During Rotation

The vestibular apparatus consists of three semicircular canals, plus the utricle and saccule. The semicircular canals resemble three hollow rings, each in a different plane approximately perpendicular to the other two (figure 6–16). They function to detect rotation, or, more properly, rotational acceleration. Each canal is filled with fluid (endolymph). In each canal, an enlarged region (the ampulla) contains a sensory structure consisting of receptors (hair cells) whose sensory hairs protrude into a gelatinous cap (the cupola). If the head rotates in the plane of the canal, the enclosed fluid, having inertia, lags behind the canal and bends the hairs in a direction opposite to the movement. For example, if the head rotates to the right (clockwise as viewed from above), the hairs in the horizontal canals are bent in a counterclockwise direction. It may help to think of the hairs as passengers in a whirling ring at an amusement park: rotation of the ring

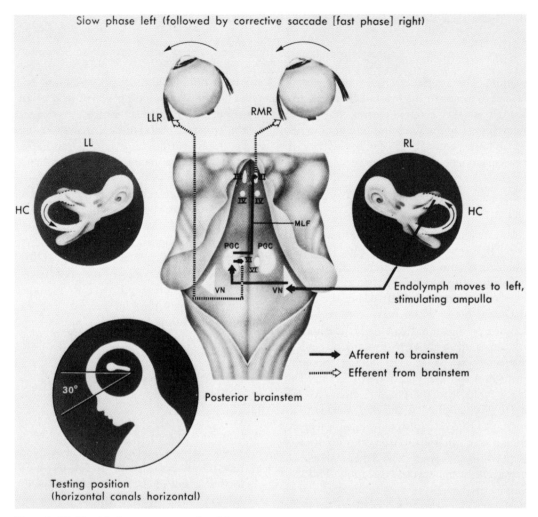

Slow phase left (followed by corrective saccade [fast phase] right)

LLR   RMR

LL   RL

HC   HC

MLF

PGC   PGC

VN   VN   Endolymph moves to left,
stimulating ampulla

→ Afferent to brainstem
┈┈⇢ Efferent from brainstem

Posterior brainstem

30°

Testing position
(horizontal canals horizontal)

Figure 6–17. Physiologic nystagmus produced by rotating the head to the right with the head bent forward 30 degrees so that the horizontal canals (HC) are parallel to the floor. Dorsal view of posterior brainstem, semicircular canals, and eyes. Rotating head to the right causes endolymph due to inertia to move to the left, stimulating vestibular nerve fibers in the ampulla of the right horizontal canal. Vestibular nerve fibers enter the brainstem to synapse in the vestibular nucleus (VN), which projects to the nuclei of cranial nerves III, IV, and VI. Efferent output via the cranial nerves to eye muscles (left lateral rectus, LLR, and right medial rectus, RMR) causes slow movement of eye to the left; a fast movement (saccade) to the right follows. The pontine gaze center (PGC) influences these eye movements, and impulses between nuclei are carried by a fiber bundle, the medial longitudinal fasciculus (MLF). Source: Gay, A.J., Newman, N.M., Keltner, J.L., and Stroud, M.H. Eye Movement Disorders. St. Louis: The C.V. Mosby Co., 1974. Reprinted with permission.

in one direction throws them backward in the opposite direction.

The vestibular nerve fibers, which innervate the hair cells, have a steady resting discharge rate. Bending the hair cells in one direction increases the rate, while bending them in the opposite direction decreases it. Rotating the head to the right (figure 6–17) excites the right vestibular nerve fibers. This causes reflex eye movements called physiological *nystagmus*. First, the eyes move slowly in a direction opposite to the rotation, as if to compensate for the head movement. This *slow phase* of nystagmus is due to connections from the vestibular nuclei in the medulla to the eye-muscle nuclei (cranial nerve nuclei III, IV, and VI) higher in the brainstem. Second, the eyes flick back quickly in the same direction as the

83

rotation. This *fast phase* of nystagmus appears to be due to inputs to the eye-muscle nuclei from sources in the cerebral cortex or reticular formation or both. With loss of consciousness, which lowers cortical and reticular activity, the fast component of nystagmus is reduced.

Clincal Testing

Clinical testing of the function of the semicircular canals and the vestibular pathways can be tested in two ways: by rotation, and by thermal stimulation (the caloric test).

Rotating the patient in a chair stimulates the canals on the right and left side of the head at the same time. If the head is tilted slightly forward, as shown in figure 6–17, the horizontal canals are parallel to the ground and are maximally stimulated. The normal response is nystagmus and the sensation of spinning. During a rotation to the patient's right, his eyes move slowly to the left and then quickly to the right; this back and forth movement is repeated during the rotation. After the rotation is stopped, for a short time there should be nystagmus in the opposite direction. This *post-rotatory* nystagmus is due to inertia: just as passengers in a ring rotated clockwise would be flung clockwise after the rotation stops, so the sensory hairs in the semicircular canals react in a similar way.

Thermal stimulation allows one to test the semicircular canals on the right and left sides separately. With the head tilted backward so that the horizontal canals are in a vertical plane, warm water is squirted into the external auditory meatus. This warms the endolymph at the outer edge of the horizontal canal on that side, and the warmed endolymph rises, flows around the canal, and bends the sensory hairs. If the vestibular system is functioning normally, this unusual stimulus gives rise to nystagmus.

The Saccule and Utricle During Linear Acceleration

Like the semicircular canals, the saccule and utricle are fluid-filled structures containing hair cells, and the hairs are imbedded in a gelatinous structure. If the head is tilted, gravity acts on this structure and causes the hairs to bend, and a specific pattern of activity is transmitted through vestibular nerve fibers to the brain. The general stimulus to these sensory organs is linear acceleration, of which gravity is one form. Other forms would occur during acceleration or braking of a car, rising and falling in an elevator, or any sudden movement of the body forward, backward, up, down, right, or left. Automatic, reflex responses of the neck, trunk, arms, and legs allow a person to maintain equilibrium during these movements. An example is the leaning forward of a runner as he accelerates during a race. The vestibulospinal tract is involved in these reflexes.

Review Exercises

1. Using a diagram of sound waves, define sine wave, amplitude, and frequency. Note how they correspond to pure tones, loudness, and pitch. Also, define harmonics and noise.
2. Explain the three functions of the middle ear.
3. Compare bone conduction and air (or ossicular) conduction.
4. Prepare a schematic diagram of the ear, indicating the tympanic membrane, ossicles, oval window, round window, cochlea, basilar membrane, organ of Corti, and helicotrema.
5. Describe how sound frequency corresponds to the location of envelope peaks of the travelling waves.
6. On the diagram of the ear, indicate where and how transduction occurs.
7. List the major structures of the auditory pathway in order, and group them according to major divisions of the brain (brainstem, thalamus, cortex).
8. Describe auditory nerve-fiber responses,

using the terms *best frequency* and *tuning curve*.

9. Define phase locking of auditory nerve fibers.

10. Describe two mechanisms of sound localization.

11. Explain why a unilateral lesion in the auditory cortex does not lead to complete deafness in the contralateral ear.

12. What is the frequency range of normal hearing?

13. Define threshold of hearing. How does it differ across frequencies?

14. Define dB.

15. Describe an audiogram.

16. Compare middle-ear deafness and sensorineural deafness.

17. Describe a brainstem auditory-evoked potential and its clinical use.

18. Describe the normal stimulus to the semicircular canals. Relate it to physiological nystagmus.

19. Give two methods for clinical testing of semicircular canal functions.

20. Describe the normal stimulus to the saccule and utricle.

21. Describe a possible common origin for the following combination of signs and symptoms. A patient complains of a hearing difficulty in the right ear and a spinning sensation and shows a slight nystagmus. A caloric test shows normal vestibular responses on the left side but reduced responses on the right side. An audiometric examination shows sensorineural deafness in the right ear. A neurological examination reveals no other abnormalities.

7. Vision

How the eye works and how the visual system converts light into sensed visual images

We are given tiny distorted upside-down images in the eyes, and we see separate solid objects in surrounding space. From the patterns of stimulation on the retinas we perceive the world of objects, and this is nothing short of a miracle [12].

Vision requires a combination of light in the outside world with the human eye and brain. If you would like to try an experiment, look with your right eye only (by covering the left with your hand) at the X in figure 7–1. This diagram now represents part of the outside world as it impinges upon your right eye, or the *visual field* of your right eye. The X, on which you are fixing your eye (or staring at), is your *fixation point*. The circle is on the right side of your visual field; although you are aware of it, it will be less distinct than the X. Now if you move the book so that it is about 20 cm from your eye, the circle should seem to disappear. At that point, it is located in the *blind spot* of your right eye.

The disappearance of the circle can be explained by considering the perspective of an outside observer looking down on you from above your head and able to see into your eye (figure 7–2). The observer would see that the fixation point (the X) is focused on a tiny area called the *fovea* in back of the eye. The circle, which is to the right of the X in the visual field, is focused to the left of the X in the eye. When the book is at the correct distance, the circle falls upon a region of the eye at the exit of the optic nerve. This region has no receptor cells to convert light into electrochemical potentials. Therefore, no information concerning the circle can be sent to the visual area of the brain, and the circle seems to disappear. This disappearance, like other aspects of vision, was not due to the visual stimulus alone, but to its interaction with the eye and brain.

In the first stage of this process, the eye focuses light from its visual field onto its back surface. In doing so the eye resembles a camera or any other optical system.

The Eye as an Optical System

When the eye is viewed from the front, its most prominent feature is the doughnut-shaped iris, pigmented gray, blue, green, or brown (figure 7–3). The hole in the doughnut is the dark-colored pupil. The pupil is really an adjustable opening through which light passes into the eye. In front of the iris and pupil is the transparent cornea, surrounded by the white sclera with overlying conjunctiva.

When the eye is viewed in horizontal section (figure 7–4), we can see the path traveled by rays of light from an object in the visual field. The light rays pass through the transparent cornea, the anterior chamber filled with a clear fluid called aqueous humor, the lens, and the transparent, gelatinous vitreous humor, and are finally focused onto the retina at the posterior surface of the eye. Befor entering the lens, the rays must pass through the pupillary opening, the size of which is controlled by the iris.

Figure 7–1. Diagram for finding your blind spot.

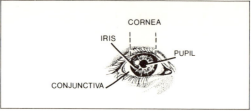

Figure 7–3. Right eye viewed from the front.

Refraction

In a camera (figure 7–5), the lens focuses light onto the film. In the eye, the optical system, composed of cornea–aqueous humor–lens–vitreous humor, focuses light onto the retina. In both cases, focusing of light is based on refraction.

A simple example of refraction is found in a triangular glass prism (figure 7–6A). As light waves enter the glass, they slow down and also bend. This slowing and bending of the light is called *refraction.* Now consider two prisms with a rectangular piece of glass in between arranged as

in figure 7–6B. The top prism bends the light downward, the middle section lets the light pass through without bending, and the bottom prism bends the light upward. At a certain distance past the prisms the light waves come together and intersect.

Figure 7–4. Right eye as viewed from above, in horizontal section. The x and the arrow are objects in the outside world (visual field) focused on the fovea and nearby retina, respectively. As in a camera, the image is reversed. Not drawn to scale.

Figure 7–2. Right eye as viewed from above, in horizontal section, while fixated on **x.** *When the circle is focused on the region shown it cannot be seen. Simplified diagram is not drawn to scale.*

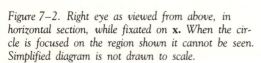

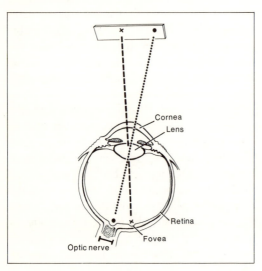

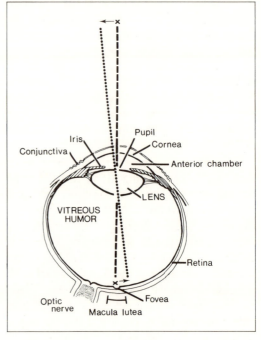

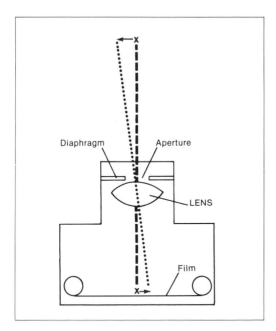

Figure 7–5. A camera viewed from above. Its lens focuses light into the film, just as the optical system of the eye focuses light onto the retina (compare with figure 7–4).

Refractive power (diopters) = 1/f (meters)

The refraction itself occurs only at the interface between two different substances. In the case of the simple lens, the first refractive surface is the interface between air and the front glass surface, and the second is the interface between the rear glass surface and air. In a camera, the

Figure 7–6. Prisms illustrate refraction. (A) Triangular prism. (B). Three prisms combined. (C) Biconvex lens.

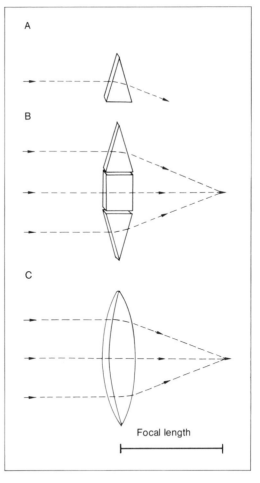

The three-prism arrangement can be considered a crude first step toward developing a biconvex lens (figure 7–6C). In the lens, as in the prisms, the top section bends the light rays downward, the center section allows them to pass straight through, and the bottom section bends them upward. In the lens, the light rays come together, or *converge*, to form a point of light at a certain distance from the lens. If the light originates at a distant point, the distance from the lens at which the light converges is called the *focal length* of the lens. The greater the ability of the lens to bend light (its refractive power), the shorter its focal length. In fact, the refractive power (in diopters) is equal to $1/f$, where f is the focal length in meters:

distance between the lens and the film surface is adjusted to equal the focal length when distant objects are being photographed. The object is then focused on the film. Each point of light in the object forms (ideally) a corresponding point of light on the film, and these points on the film together constitute an image of the object. Compared to the object, the image is reversed in direction and upside-down.

A good camera contains a compound lens, composed of several lenses cemented together and thus several refractive surfaces. Similarly, the eye has several refractive surfaces as follows:

air / cornea / aqueous humor / lens / vitreous humor

Most of the refraction actually takes place at the first refractive surface, between air and the anterior of the cornea. The major contribution of the lens is to provide adjustable refraction.

A distant object (say, a star or a penlight from 50 feet away) will form a point focus on the retina, approximately 24 mm behind the cornea of a normal, relaxed eye. A simplified optical model of such an eye has a focal length of about 17 mm; its refractive power is 1/0.017 meters or approximately 59 diopters.

Refractive Errors

Difficulty in clearly focusing images can be due to three major sources of refractive error: myopia, hyperopia, and astigmatism.

In *myopia* (figure 7–7A), distant objects form an image in front of the retina; at the retina itself, the image is blurred. People with myopia are also called nearsighted because they have difficulty seeing distant objects clearly, while nearby objects can often be seen very well. Myopia arises because the distance between cornea and retina is too long for the refractive power of the eye. It can be corrected with a concave lens, which diverges (spreads

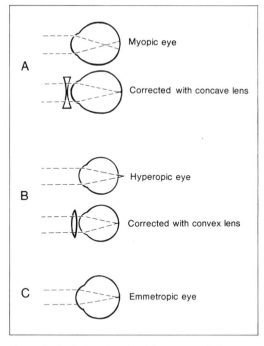

Figure 7–7. Common refractive errors and their correction. (A) Myopic eye. (B) Hyperopic eye. (C) Emmetropic eye (normal refraction).

apart) the light waves and allows them to come to a focus on the retina.

In *hyperopia* (figure 7–7B), distant objects from an image behind the retina; again, at the retina itself, the image is blurred. Hyperopia may be called farsightedness because the problem is worse for near than for faraway objects. It arises because the eye is too short for the refractive power of the lens. Hyperopia can be corrected with a convex lens, which converges the light waves and allows them to come to a focus on the retina.

In *astigmatism*, the curvature of the cornea (or other refractive structure) may be greater in one direction (e.g., vertically) than in another (e.g., horizontally). The image of a point then becomes an ellipse on the retina. In the example given, the problem can be corrected with a lens that is more curved horizontally than vertically.

Accommodation

When a photographer wants to take a close-up picture, he focuses the lens by increasing the distance between the lens and the film surface. The mammalian eye works on a different principle. When an object is moved closer to the eye, the distance between cornea and retina does not change. Instead, the refractive power of the eye's optical system is increased to keep the image in focus. This process is called *accommodation,* and requires the lens to change shape—it becomes rounder, or more convex (figure 7–8). The more convex the lens, the greater its refractive power. Some intricate mechanical adjustments are involved in this process. Left to itself, the lens would be almost round when viewed in horizontal section. However, in the eye, the lens is being flattened by the pull of suspensory ligaments (zonular fibers) on its outer border (these ligaments form a ring around the lens as viewed from front or rear). Contraction of the ciliary muscle releases this tension and allows the lens to assume a more spherical shape due to its own elastic properties.

The ciliary muscle is controlled by parasympathetic fibers from cranial nerve III (oculomotor), which secrete acetylcholine at the neuromuscular junction. The accommodation response involves areas of cerebral cortex and the superior colliculus. As in other parasym-

Figure 7–8. Accommodation of the lens for near vision (right), compared with the unaccommodated eye (left). Arrows indicate contraction of ciliary muscle during accommodation.

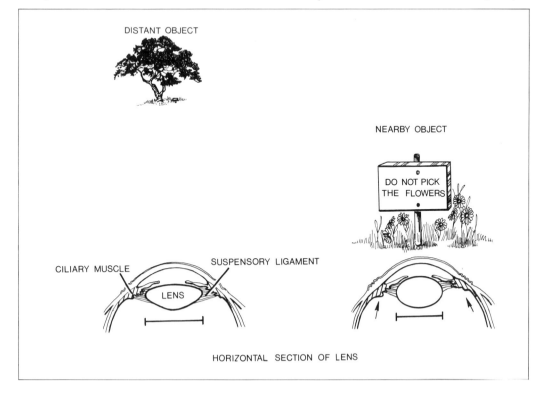

pathetic responses, accommodation can be blocked by the drug atropine.

The capacity of the lens to become more convex when tension is released is also essential for accommodation to occur. With increasing age, the lens tends to become more rigid, and accommodation is impaired. This condition is called *presbyopia*. It may be partly compensated for with bifocal lenses, in which the lower part of the spectacle lens is used for near vision.

Adjustment of Pupil Size

The amount of light entering the lens is determined by the size of the pupil, which in turn is controlled by the surrounding iris. The iris-diaphragm in a camera performs a similar function: in bright sunshine it may be adjusted to provide an opening as small as a pinhead (signified as f-stop f-22), while in dim indoor light it may be opened wide (f-2.8) to allow enough light to enter the camera to provide adequate exposure for the film. If you look at people's eyes in a wide range of lighting conditions, you may find pupillary diameters ranging from about 2 to 8 mm (figure 7–9). Although other influences are also involved, brighter light causes smaller pupils. You can try a simple test on yourself to demonstrate the pupil's response to a change in illumination. If you close your eyes for about 25 seconds and then open them while looking into a mirror in

a well-lit room, you should see both pupils constrict in response to the increased illumination.

This *constriction* in response to light is called the *pupillary light reflex*. It can be demonstrated by asking another person to look ahead in a moderately lit room and then shining a penlight into the eye. The pupillary constriction that results is an automatic reflex response that protects the retina against too-intense light, just as a small iris-diaphragm protects film against overexposure. The effector muscles are the sphincter (circular) muscles of the iris, which receive a parasympathetic, cholinergic innervation from cranial nerve III (oculomotor). The pupillary light reflex pathway involves the following:

light → optic nerve tract → pretectal nucleus → E-W nucleus → cranial nerve III → sphincter muscle of iris → constriction

where E-W stands for the Edinger-Westphal or parasympathetic nucleus of cranial nerve III, located, along with the pretectal nucleus, in the midbrain. The time required to traverse this pathway, from light onset to pupillary constriction, is approximately 200 msec. In the pupillary light reflex, both pupils constrict even though only one may be stimulated. This is due to fibers that cross over in the posterior commissure, from pretectal nucleus on one side to E-W nucleus on the other. Pupillary constriction is also called *miosis*.

Active *dilation* of the pupil is caused by

Figure 7–9. The approximate range of pupil diameter.

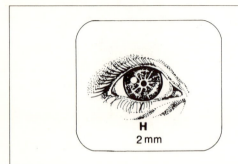
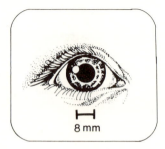

contraction of the dilator (radial) muscles of the iris, which receive innervation from sympathetic fibers that secrete norepinephrine. Via this pathway strong emotions like fear or excitement can cause pupillary dilation.

The balance between the opposing cholinergic and adrenergic effects on the iris explains how many drugs can affect pupil size (figure 7–10). Atropine or similar substances given in eyedrops block acetylcholine receptors in the sphincter muscles of the iris; the unopposed dilator muscles then cause pupillary dilation, an effect used by ophthalmologists when they examine the eye. Amphetamine also causes dilation, although by a different mechanism; it enhances the response of the dilator muscles. Acetylcholine-like substances, such as pilocarpine, activate the sphincter muscles of the iris and cause pupillary constriction. Cholinesterase inhibitors (eserine, prostigmine) allow a

buildup of acetylcholine, which also activates the sphincter muscles and causes pupillary constriction.

You may have noticed that cholinergic fibers from cranial nerve III cause two effects on the eye: accommodation for near vision and pupillary constriction. During an eye examination, the atropine that blocks pupillary constriction (and leads to a dilated pupil) will block accommodation as well. The patient will have difficulty seeing clearly, especially nearby objects, until the effects have worn away. Another occasion in which both effects are evident is the *near response:* when a person looks from a distant object to a nearby one, accommodation and pupillary constriction occur automatically. The pupillary constriction sharpens the image, because light rays are forced to pass only through the center of the lens, reducing the distorting effects (spherical and chromatic aberrations) of light passing through the edges. These aberrations would otherwise tend to worsen as the lens becomes rounder.

Figure 7–10. Opposing effects of sympathetic (adrenergic) and parasympathetic (cholinergic) innervation of the iris, and the effect of drugs. From Westheimer, G. The eye, including central nervous system control of eye movements. In Mountcastle, V.B. (Ed.). Medical Physiology (12th ed.). Vol. 2. St. Louis: The C.V. Mosby Co., 1968. Reprinted with permission.

The Neurophysiology of Vision

The ability to see light and dark differences, forms, and details (pattern vision) may, for the

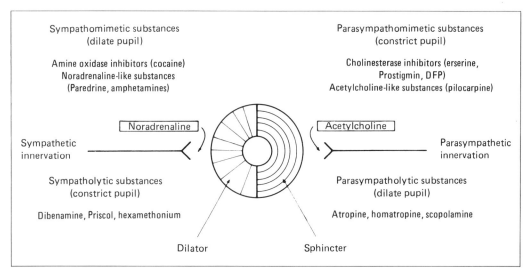

sake of simplicity, be studied as a function of input to one eye only (monocular vision) and without reference to color. Binocular and color vision will be discussed later.

Transduction in the Retina

The optical apparatus of the eye focuses on the retina an image of the visual field, reduced to the size of a postage stamp and inverted, as is the image focused on the film of a pocket camera. The varying intensities of light in this image are absorbed by the photoreceptors of the eye, called rods and cones. Specifically, the light is absorbed by a visual pigment in the photoreceptors; in the case of the rods, this pigment is called *rhodopsin*. In the presence of light, a photochemical reaction takes place (figure 7–11) in which rhodopsin breaks down to a colorless pigment, retinene, plus a protein,

Figure 7–11. The photochemical reaction occurring in the rods. Lumirhodopsin and metarhodopsin are intermediate chemical forms produced by the action of light on rhodopsin. Source: Reprinted with permission from Brown, K.T., Physiology of the retina. In Mountcastle, V.B. (Ed.), Medical Physiology (12th ed.). Vol. 2. St. Louis: The C.V. Mosby Co., 1968. Adapted from Wald, G., in Magoun, H.W. (Ed.), Handbook of Physiology. Baltimore: Williams & Wilkins, 1959.

opsin. The retinene is enzymatically reduced to vitamin A. The photochemical reaction causes a visible bleaching of the deep purple rhodopsin to a colorless form. At the same time, the reaction gives rise to a transduction from light to electrical energy: a receptor potential occurs in the photoreceptors.

This receptor potential is most unusual. In most receptors, it consists of depolarization, due to an increased conductance to sodium and other small ions. In the photoreceptors, however, it consists of a *hyperpolarization*—the interior of the cell becomes more negative—due to a decreased conductance to sodium. This hyperpolarization develops in a cell that, in comparison with other receptors, is already partially depolarized, having a resting potential of -20 to -40 mV. The amount of hyperpolarization depends on the intensity of light: a weak light stimulus might cause a hyperpolarization of only 2 mV (from -20 to -22 mV), while a brighter stimulus might cause a hyperpolarization of 15 mV (from -20 to -35 mV).

Synaptic Actions in the Retina

The rods and cones synapse with two other cell types: horizontal cells and bipolar cells (figure 7–12). The presence of vesicles in the presynaptic region suggests that a release (or reduced

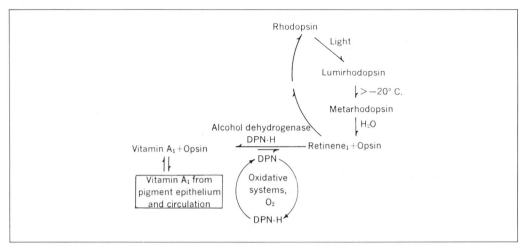

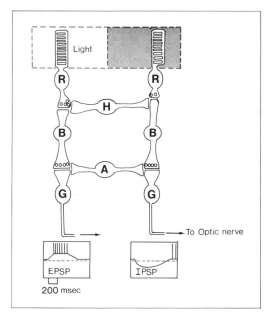

Figure 7–12. General scheme of cell types and synaptic connections in the retina. **R** = *rods,* **B** = *bipolar cells,* **G** = *ganglion cells,* **H** = *horizontal cells, and* **A** = *amacrine cells. Light falling on the receptive components of the rod causes synaptic and spike potentials in ganglion cells as shown below. After Dowling [6].*

release) of a chemical transmitter is involved in the transmission of information from the photoreceptors. The bipolar cells relay information to the ganglion cells, which in turn send information in the form of nerve impulses along their axons, which bundled together make up the optic nerve. Sideways transmission of information is handled by the horizontal cells, which connect adjacent photoreceptors and bipolar cells, and by the amacrine cells, which connect adjacent bipolar cells and ganglion cells. Thus, the basic circuit of the retina can be visualized as two rungs of a ladder, with the sides composed of the sequence of photoreceptor-bipolar-ganglion cells and the rungs composed of horizontal cells and amacrine cells.

With this simplified model in mind, we can see how a single illuminated photoreceptor could send information via a bipolar cell to a ganglion cell, which would respond with a depolarization (an EPSP) and a series of action potentials.

At the same time, the adjacent ganglion cell could respond with a hyperpolarization (an IPSP) and a cessation of action potentials. The inhibition of the adjacent ganglion cell could be due to an inhibitory action of one or both of the laterally–transmitting cells (the horizontal and amacrine cells), and is an example of lateral inhibition, similar to the mechanism in somatic sensation. In brief, the outcome would be that illumination of one receptor would cause excitation of the directly connected ganglion cell and inhibition of surrounding ganglion cells. If we record from a single ganglion cell while moving the light stimulus, the ganglion cell would be excited by illumination of a directly connected receptor and inhibited by illumination of a neighboring receptor.

In reality, the ladder model is an oversimplification; a number of photoreceptors can converge to synapse on a single bipolar cell, and a number of bipolar cells can converge on a single ganglion cell. In addition, most stimuli illuminate more than one photoreceptor. This leads to the more general result of a retinal ganglion cell being excited by illumination of its more directly connected receptors and inhibited by illumination of surrounding receptors.

Experimentally, most studies of ganglion cells have been done one cell at a time, while illuminating various spots on the retina. For a single ganglion cell, illumination of only a small circular area of the retina, generally overlying the cell, causes either excitation or inhibition; this area is called its *receptive field* (figure 7–13). In the visual system as a whole, the receptive field of a neuron is "the area of retina from which the discharges of that neuron can be influenced" [6]; this area of retina corresponds to a particular area of the visual field. Stimuli outside the receptive field have

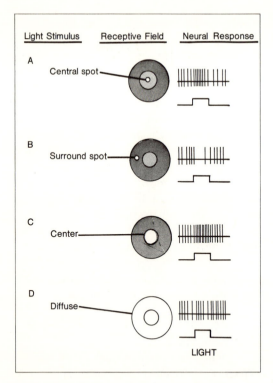

Figure 7–13. Receptive field of a retinal ganglion cell with excitatory center and inhibitory surround. Neural spike responses shown on right.

Figure 7–14. Tests for visual acuity: (a) Landolt–C; (b) and (c) two examples of letters from Snellen test. The gap or space, s, subtends a known angle at the eye. Source: Davson, H., and Eggleton, M.G. Principles of Human Physiology (14th ed.). Philadelphia: Lea & Febiger, 1968. Reprinted with permission.

no effect, although the cell may continue discharging spontaneously. For a ganglion cell, illumination confined to the center of the receptive field (presumably stimulating the more directly connected photoreceptors) may cause excitation (figure 7–13A and C) while illumination confined to the concentric surround area may cause inhibition (the firing rate may be reduced below the spontaneous level; figure 7–13B). Thus the receptive field is composed of an excitatory center and an inhibitory surround. In other retinal ganglion cells, the receptive field is reversed, with an inhibitory center and an excitatory surround. When the entire receptive field is illuminated (figure 7–13D), the surround response reduces the center response.

Visual Acuity and Retinal Receptive Fields

Visual acuity—the ability to see detail—can be measured by the Landolt C test, which tests the ability to see the small gaps in the rings (figure 7–14a). The person stands a fixed distance from the chart and is asked to indicate the direction of the gaps in successive characters. For measuring acuity, the size of the gaps is translated into a visual angle. Under good conditions, a person should be able to detect a gap that subtends an angle of one minute of arc (1/60 of a degree). The gap in a small printed c in this text subtends such an angle at a distance of approximately 3.4 meters. The familiar Snel-

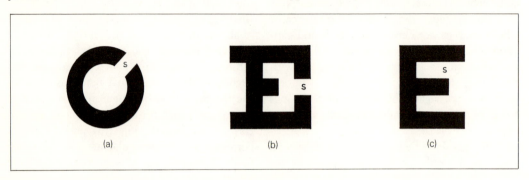

len Chart, with a series of letters on each line, is used in a similar manner: a person with normal vision, standing 20 feet from the chart, can just recognize the letters on the line marked 20 feet. The gaps in the letters (figure 7–14b and 7–14c) will then subtend a visual angle of one minute; and the person will be said to have a visual acuity of 20/20. These tests are commonly given to test the optical function of the eye but may also be used to test the visual pathway.

Visual acuity can be compared with two-point tactile discrimination. We have seen that the two-point threshold is smallest on parts of the body (fingertips, tongue) having small receptive fields and a high density of receptors per square centimeter. Similarly, visual acuity varies over different parts of the visual field and, thus, over corresponding parts of the retina, and the variation can be related to changes in retinal receptive fields.

The variation in acuity can be illustrated by fixating a single letter on the printed page with one eye covered; letters at increasing distances to either side will be more difficult to recognize. The principle illustrated is that visual acuity is good at the fixation point but falls off rapidly toward the periphery. This relationship is illustrated in the graph of figure 7–15.

Visual acuity is related to the size of the receptive-field centers of retinal ganglion cells: small receptive-field centers are required to see fine detail. The smallest receptive-field centers occur in the foveal region and consist of one to three photoreceptors only. With increased distance from the fovea, the size of the receptive-field centers grows to several degrees, as a number of photoreceptors converge to activate a single ganglion cell.

Edge Enhancement and Retinal Receptive-Field Organization

The organization of ganglion-cell receptive fields into an excitatory center and an inhibi-

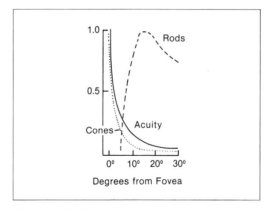

Figure 7–15. Graph of visual acuity (solid line), against visual angle (from the fixation point). Cone concentration (dotted line) and rod concentration (dashed line) plotted against distance from fovea in degrees. Vertical axis shows relative units only.

tory surround is an example of lateral inhibition. As in somatic sensation, it causes an exaggerated neural response to edges. In vision, the edge effect can be seen in a series of strips like those prepared by photographers testing exposure time in the darkroom (figure 7–16). Although each strip really consists of a constant light intensity, subjectively, it appears as if the darker strip becomes darker near its border with the lighter strip, and the lighter strip becomes lighter near its border with the darker strip. This can be explained by noting that the ganglion cell with a receptive field completely illuminated by the lighter strip has a slightly lower firing rate, and therefore leads to less perceived light intensity, than the ganglion cell having less of its inhibitory surround illuminated due to the impingement of the dark border.

Rod and Cone Vision

The two kinds of photoreceptors, rods and cones, differ in several ways.

Distribution. The fovea itself contains only cones. With distance from the fovea, as measured in terms of visual angle (degrees from

Figure 7–16. Strips of stepwise increasing light intensity, demonstrating accentuation of edges (also called the Mach band effect). Superimposed receptive fields of retinal ganglion cells illustrate a possible mechanism for this effect. Receptive field on left has less illumination of its inhibitory surround (minuses), therefore its cell fires at higher rate than cell on right.

fovea), the concentration of cones per unit area declines, and up to a point, the concentration of rods increases (figure 7–15). In the periphery of the retina, the photoreceptors are only rods.

Light sensitivity. Vision in moderately bright light is a product of both rods and cones. Vision in very dim light is due only to rods and requires a substantial delay for dark adaptation to occur. A good example of dark adaptation occurs when you walk from a bright lobby into a dark movie theater. Initially you can see very little and only gradually does light sensitivity increase enough to allow you to see well. When this process is measured in the laboratory, it turns out that the time-course of the adaptation is composed of two curves: the first reaches a plateau in 10 minutes and is due to dark adaptation of the cones, and the second reaches a plateau at a greater level of light sensitivity in about 40 minutes and is due to adaptation of the rods. After this 40-minute delay, vision in dim light is essentially rod vision.

The astronomer Kepler made an interesting observation that can be related to both the distribution and light sensitivity of the rods. He found that a dim star could be seen only if viewed obliquely, a short distance from the fixation point. This was later attributed to the ability of rods rather than cones to respond to a dim object, and to the presence of rods only a short distance from the fovea rather than within the fovea itself.

Dark adaptation is due in part to the resynthesis of visual pigment (figure 7–11). In the dark, the retinene and opsin in the rods become converted into light-sensitive rhodopsin (in the cones a similar process leads to synthesis of the cone pigment, photopsin). In addition, large quantities of vitamin A are converted to retinene, which is then available for resynthesis of additional visual pigment. When the photoreceptors contain a higher concentration of rhodopsin and photopsin due to this resynthesis, they are more sensitive to light. Without vitamin A, light sensitivity is poor, particularly in dim light, giving rise to night blindness.

Color sensitivity. The color vision we have in bright light is due to the cones. Rod vision allows us to see black, white, and shades of gray, as we do in very dim light.

In addition, the rods and cones respond to overlapping but not identical parts of the color spectrum: the rods do not respond to pure red light (the longest wavelengths in the visible spectrum), while the cones do. Thus, a pure red light stimulus is a handy method for experimentally activating cones without rods. In one practical application, night-flying pilots can avoid the 40-minute wait for dark adaptation by wearing red goggles for some time before takeoff. Since only red light passes through the goggles to the eyes, the rods are not activated and adapt just as they would in darkness. After removing the goggles, the pilot can immediately see in the dim light of the night sky using rod vision.

Visual acuity. The greatest visual acuity (the ability to resolve details separated by a small visual angle) is due to cone vision and is found

Table 7–1. Characteristics of Rod and Cone Vision

Characteristics	Measurement	Cones	Rods
Distribution (in retina)	Distance from fovea as measured in degrees of visual angle	High concentration in fovea; concentration declines toward periphery of retina	Absent in fovea; predominant in periphery of retina
Light sensitivity	Minimum intensity of light stimulus causing response (sensitivity is the inverse of this value)	Low sensitivity	High sensitivity (can see in dim light after dark adaptation)
Color sensitivity	Wavelength	Color vision	Black-white vision; also, insensitive to pure red light
Visual acuity	Minimum visual angle between objects (acuity is the inverse)	High acuity (can see fine details)	Low acuity

in the immediate vicinity of the fixation point, which is focused on the all-cone fovea. As figure 7–15 shows, visual acuity falls off with distance from the fovea in the same manner as does cone concentration. Rod vision is capable of much less acuity than is cone vision. Thus, we are much more capable of distinguishing details, such as fine print, in bright light when fixating directly on the object, than in dim light or in the periphery of the visual field, where rod vision predominates.

In summary, as shown in table 7–1, rod and cone vision differ in several characteristics. Rod vision is capable of high sensitivity to light (it responds to low light intensity) but cannot distinguish colors (wavelength) and has poor visual acuity (it distinguishes between points only if they are separated by a large visual angle). Cone vision has only low sensitivity but is capable of color vision and high acuity in bright light.

The Visual Pathway in the Central Nervous System

In each eye, the axons of the approximately one million retinal ganglion cells come together to form the *optic nerve* (figure 7–17). About half of these cells represent the left half of the retina (left hemiretina) and the other

half represent the right half of the retina (right hemiretina). The two optic nerves meet at the *optic chiasm,* where a regrouping of nerve fibers occurs. This regrouping is most easily understood in terms of the visual field, which can be divided into a left and right half at the fixation point. For each eye, the left half of the visual field is focused onto the right half of the retina. The corresponding optic nerve fibers combine at the optic chiasm to form the right *optic tract.* Thus, at the optic chiasm, uncrossed fibers from the right hemiretina of the right eye are joined by crossed fibers from the right hemiretina of the left eye. These fibers carry information representing the left half of the visual field to the visual pathway of the right side of the brain. Similarly, for each eye, the right half of the visual field is focused onto the left half of the retina. The corresponding nerve fibers form the left optic tract, which consists of uncrossed fibers from the left eye and crossed fibers from the right eye. These fibers carry information to the visual pathway of the left side of the brain.

The main visual pathway on each side consists of the *optic tract,* the *lateral geniculate nucleus,* the *optic radiation,* and the *primary visual cortex* (Brodmann's area 17). A large lesion in this pathway can cause blindness in the opposite half of the visual field of both eyes, a condition termed *hemianopia*. Nerve fibers in the optic tract synapse onto cells in the lateral geniculate, which is a relay nucleus in the thalamus. The fibers from these cells form the optic radiation, which projects to cells in the primary visual cortex, in the occipital lobe. The retinal surface is mapped onto the visual cortex, just as the body surface is mapped onto the somatic sensory cortex. The heavy concentration of incoming nerve fibers from the optic radiation in layer IV of the six-layered visual cortex forms a horizontal striation, which gives rise to another name for this region, the striate cortex. Outgoing fibers leave the striate cortex for the adjacent visual association cortex (areas 18 and 19).

Neuron Activity in the Visual Cortex

The circular organization (center-surround receptive field) of retinal ganglion cells is preserved, with slight change, in the lateral geniculate nucleus and in some of the cells

*Figure 7–17. Schematic diagram of the visual pathway as viewed from above the head. **T** = temporal, **N** = nasal half of visual field; **L** = left, **R** = right.*

located in layer IV of the primary visual cortex near the incoming nerve endings from the lateral geniculate. Other cortical cells, however, have a new type of receptive-field organization. These cells, studied extensively by Hubel and Wiesel [17,18], require not only that the stimulus be located in a specific area of the visual field (as do retinal ganglion cells), but also that, within that area, excitatory and inhibitory areas are arranged along a line at a particular angle from the vertical (a particular orientation). As shown in figure 7–18, the optimal stimulus is a bar of light at an angle that covers the excitatory but not the inhibitory receptive field. Rotation of the bar reduces the neural response, since both excitatory and inhibitory areas are activated; a bar that is oriented perpendicular to the correct angle evokes little or no response (figure 7–18D). These cells may be called orientation-specific edge detectors; Hubel and Wiesel called them simple cells to contrast with other, more-complex cells in the visual area.

What is the basis for the response specificity of the simple cortical cells? One possibility, suggested by Hubel and Wiesel, is that these

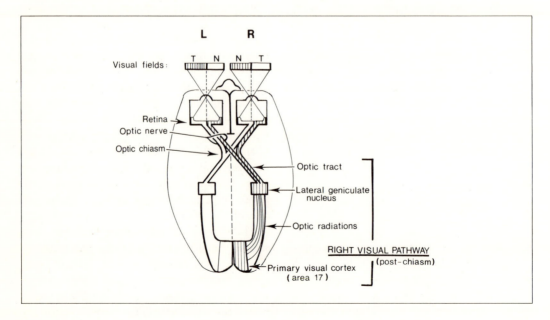

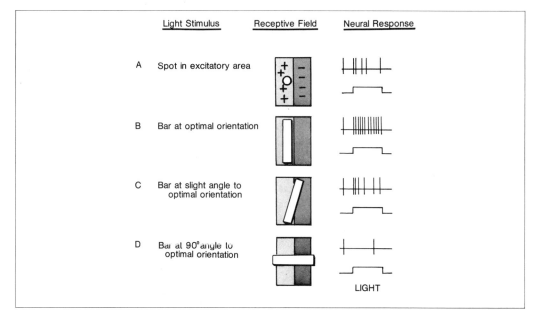

Figure 7–18. Receptive field of a neuron in visual cortex. This receptive field responds maximally to a vertical bar. + = excitatory area, − = inhibitory area. After Hubel and Wiesel [16].

cells receive the convergent input of several neurons in the lateral geniculate nucleus or in layer IV of the striate cortex, each with circular receptive fields adjacent to each other along a single line in the visual field (figure 7–19). The compound receptive field composed of the adjacent circular receptive fields would have the orientation specificity of the simple cortical cells.

The more-complex cortical cells are of several types. Some have orientation specificity but less location specificity than simple cells— they will respond when the bar or edge is shifted in position over a moderate range. Others respond best to a corner at a particular angle, moving in a particular direction.

We have noted that the visual cortex (like other primary sensory-cortical areas) may be divided into six cell layers. After studying the responses of cells encountered by microelectrodes passing through all six layers at different angles from the cortical surface, Hubel and

Wiesel presented a model of how the receptive fields of cortical cells are organized [17,18]. A simplified version of this model appears in figure 7–20. A microelectrode lowered perpen-

*Figure 7–19. Schematic model explaining receptive field of a simple visual cortex cell. Several cells in the lateral geniculate nucleus (**LGN**), with receptive fields lined up vertically in the retina, synapse onto a single cortical cell as shown (**e** = excitatory synapse). After Hubel and Wiesel [16]. Source: Shepard, G.M., The Synaptic Organization of the Brain (2nd ed.). New York: Oxford University Press, 1979. Reprinted with permission.*

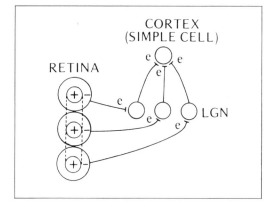

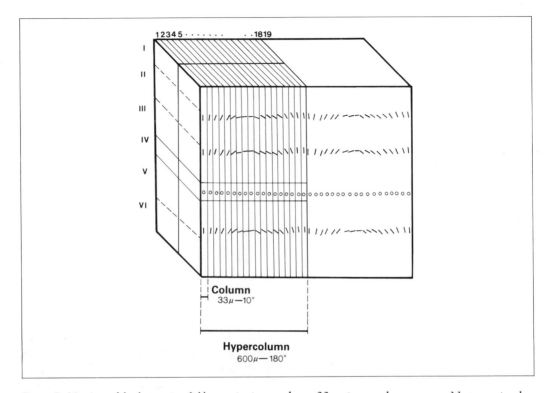

Figure 7–20. A model of receptive field organization in visual cortex. I to VI represent the layers of cerebral cortex (I is the surface layer). Short lines and circles represent receptive fields of neurons in the cortex as described in text. Numerical measurements given are representative only. Adapted from D.M. Hubel and T.N. Wiesel. Brain mechanisms of vision. Sci. Am. 241:150–162, 1979.

dicular to the cortical surface might encounter a number of cells whose receptive field covers approximately the same area of the visual field. Some cells in layer IV, as we have mentioned, would have circular receptive fields, as indicated by the small circles in figure 7–20. However, other neurons encountered by the microelectrode would have vertically oriented receptive fields as symbolized by the short vertical lines in figure 7–20. These neurons could be visualized as a slab, or orientation column, of cells with similar response properties, and are labeled column 1 in the figure. The width of the column was found to average

about 33 μ in monkey cortex. Neurons in the adjacent column, labeled column 2, have receptive fields with preferred orientations about 10 degrees clockwise from the vertical, neurons in column 3 have fields with orientations about 20 degrees clockwise, and so on. With each column covering about 10 degrees, all possible angles (180 degrees) could be covered by 18 orientation columns, occupying a width of about 600 μ in the cortex. The 600 μ-wide hypercolumn would be able to respond to all line orientations within a particular receptive field, covering a certain number of minutes or degrees of visual angle. The next hypercolumn, starting with column 19, would represent an immediately adjacent area in the visual field. Figure 7–21 shows typical relationships between two hypercolumns, labeled *a* and *b*, and their corresponding receptive fields. Note that the adjacent hypercolumns correspond to adjacent and partially overlapping receptive fields. The upshot of this model is that within a block

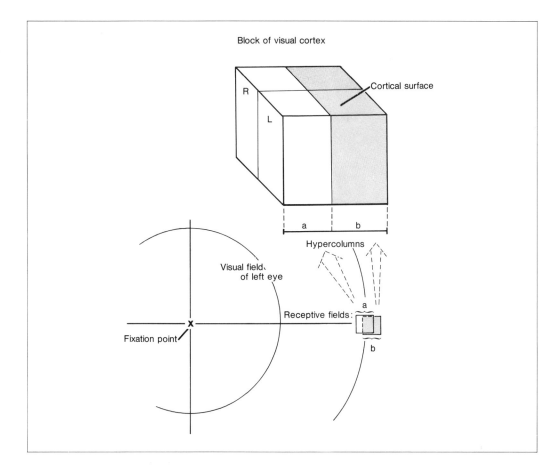

Figure 7–21. Projection of receptive fields (below) onto hypercolumns of visual cortex (above) in the model suggested by Hubel and Wiesel [17,18].

of cortex approximately 600 μ (one hypercolumn) by 2,000 μ (the depth of the cortex) is the neural machinery for analyzing all possible light-dark borders or lines within a given area of the visual field. An adjacent block of cortex analyzes an adjacent area of the visual field. Within each block is a similar series of orientation columns.

Binocular Vision

By alternately closing the left and right eyes, you can alternate between right and left mon-

ocular vision and produce a shift of the visual image especially for nearby objects. With both eyes fixed on the same object, the resulting binocular vision involves a fusion effect, in which the object is perceived as being in a single location. This binocular fusion results when the image of the object falls on corresponding areas of both left and right retinas.

Neurons carrying information about these two corresponding areas remain essentially separate all along the visual pathway until the striate cortex. Even in the striate cortex, a number of neurons are monocular—they have receptive fields in one eye only. The model in figure 7–21 shows how right or left eye dominance of neural responses appears to alternate at right angles to the sequence of orientation

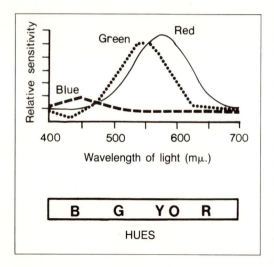

Figure 7–22. *Sensitivity curves of blue-, green-, and red-sensitive cones.* **B** = *blue,* **G** = *green,* **Y** = *yellow,* **O** = *orange, and* **R** = *red. Adapted from R.L. Gregory.* Eye and Brain: The Psychology of Seeing *(3rd ed.). New York: World University Library, McGraw-Hill Book Co., 1977.*

columns. Two of these ocular dominance areas, representing both right and left eyes, can be accommodated within about 800 μ. In addition to monocular neurons, a large number of orientation-specific neurons are binocular. Although they can be excited by either eye separately, they are excited to a greater extent by stimulation of approximately corresponding locations on both retinas. The behavior of these neurons may provide a physiological basis for binocular vision.

Color Vision

In dim light most vision is by means of the rods and involves little or no color vision. In bright light, cone function makes possible color vision. Three different kinds of cones, each containing a pigment most sensitive to light of a particular color, have been found in the human retina (figure 7–22). The blue-sensitive cones have maximal sensitivity in the short-wavelength region (455 millimicrons); the green-sensitive cones have maximal sensitivity in the moderate-wavelength region (535 mμ); and the red-sensitive cones have maximal sensitivity in the long-wavelength region (575 mμ). Light stimuli often activate two or more of these cone types at once. For example, a yellow light of 560 mμ will activate both green and red cones; the corresponding neural activity leads to the perception of yellow. If pure green and pure red light are made to fall upon the same area of the retina (this can be done, for example, by projecting a green and red spotlight on the same part of a white screen) a yellow color will result. Such additive combination of colors should not be confused with the subtractive combination that occurs when paints are mixed.

Various forms of partial color blindness are found in a small portion of the population. One form, called *protanopia,* is explained by a lack of red cones; another form, called *deuteranopia,* is explained by a lack of green cones. Both may be associated with the tendency to confuse red and green. Tritanopia, explained by a lack of blue cones, is rare. Also rare is total color blindness, which can generally be explained by the presence of only the pigment rhodopsin in both rods and cones. People with this condition can perceive only various shades of gray.

The three types of cones converge onto retinal ganglion cells and give rise to several types of response characteristics. For example, in one type, the center of the receptive field is excitatory, and the surround is inhibitory, regardless of the wavelength of the light stimulus. In another type, however, the center and surround respond differently to different colors: the center may be excited by red and inhibited by green, while the surround may be excited by green and inhibited by red. A third type of receptive field is differentially sensitive to yellow and blue. These receptive-field types of retinal ganglion cells represent further refinement of color processing in the visual system.

Review Exercises

1. Draw a diagram of the eye and on it point out the refractive surfaces, the retina, the fovea, and the optic disk. Describe the relationship between the visual field, fixation point, and blind spot, with the retina, fovea, and optic disk.

2. Describe myopia, hyperopia, presbyopia, and astigmatism, and corrective lenses for each.

3. Describe the pupillary light reflex and its reflex pathway.

4. Describe accommodation.

5. List the effects of chemical transmitters and drugs on pupillary size.

6. State the relationship of vitamin A to rhodopsin.

7. Diagram the neural structure of the retina.

8. State the definition of receptive field. Describe an experimental method for plotting the receptive field of a retinal ganglion cell. Diagram a typical example.

9. Describe the basis for the clinical tests for visual acuity.

10. Differentiate rod and cone vision in terms of retinal distribution, light sensitivity, color sensitivity, and visual acuity.

11. Diagram the visual pathway, identifying optic nerve, optic chiasm, optic tract, lateral geniculate nucleus, and cortical area 17. On the diagram, show the projection of temporal and nasal halves of the visual field onto the retina and the visual cortex. Illustrate effects of lesions in an optic nerve, tract, or chiasm on visual fields.

12. Use the term *simple cell* in describing neuronal activity in the visual cortex. Describe and diagram the receptive field of such a cell.

13. State the current view of the mechanism for color vision.

14. State the current view of the mechanism for binocular vision.

15. Early astronomers discovered that a dim star was best seen if they looked slightly to one side of it rather than directly at it. How can this be explained on the basis of (1) the distribution and (2) the light sensitivity of the rods and cones in the retina?

16. After receiving atropine eye drops in the left eye only, a patient is asked to focus on a near object. Describe and explain the different responses in the two eyes.

17. A patient is found to have no vision in the left half of both visual fields. Name this field defect and suggest the site of a possible lesion or dysfunction.

8. Chemical Senses

How the nervous system responds to chemicals to produce sensations of taste and smell

Taste

Compared with touch, pain, hearing, and vision, taste may appear to be a relatively unimportant sense, a concern of cookbooks and gourmet chefs. But its role in nutrition, from the first drop of milk swallowed by the newborn infant, shows that taste is a sense worth attending to.

Structure

The *taste buds* are receptor organs located on the tongue and the walls of the pharynx. They are clustered on the surface of small projections (the fungiform, foliate, and vallate papillae) that can be seen with the unaided eye. Each taste bud contains a number of sensory-receptor cells, with projecting microvilli, whose membranes can interact with chemical solutions that enter the taste bud through a pore (figure 8–1).

Responses to appropriate stimuli are relayed via synapses on the receptor cells to afferent-nerve fibers in cranial nerves VII (facial) and IX (glossopharyngeal). These two cranial nerves serve different areas: the facial nerve innervates the anterior two-thirds of the tongue, and the glossopharyngeal nerve innervates the posterior one-third of the tongue. Both cranial nerves enter the central nervous system at the level of the medulla, where they join the solitary tract, and then synapse in the nucleus of the solitary tract. From there, a pathway mediating *taste sensation* travels ros-

trally, relaying in the pons and thalamus en route to a cortical taste area on the postcentral gyrus, adjacent to the face part of the somatic sensory area. Other pathways, mediating automatic *reflex responses* to taste stimuli, leave the nucleus of the solitary tract for nuclei within the brainstem. Such reflex responses include tongue responses to sweet and other substances, involving cranial nerve XII (hypoglossal), and the salivation response to food placed in the mouth.

Taste Sensations and Sensory Responses

There are four basic qualities of taste sensation: sweet, salt, sour, and bitter. Examples of chemicals that give rise to these qualities are shown in table 8–1. Sour taste is elicited by the hydrogen ions released by acids, and salty taste by inorganic salts, particularly those containing chloride ions. Sweet taste is elicited typically by sugars, but also can be elicited by other organic compounds, such as the artificial sweetener, saccharin, and by some inorganic substances, such as lead (whose taste may contribute to the ingestion of lead-based paint chips by children). Bitter taste is stimulated by organic substances, especially alkaloids such as quinine, nicotine, and caffeine.

Sensitivity to the four taste qualities is unevenly distributed on the tongue (figure 8–2). The tip of the tongue is sensitive to sweet substances and, together with salt- and sour-sensitive areas, is innervated by the facial nerve. The base of the tongue is sensitive to

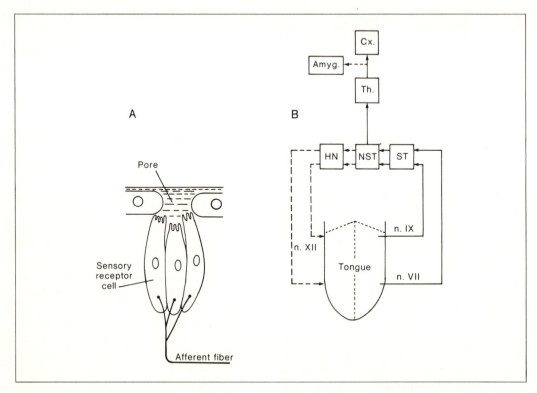

Figure 8–1. (A) A taste bud, showing several sensory receptor cells, Supporting cells are associated with receptor cells but are not shown. (B) The neural pathway for taste. Afferent fibers travel in cranial nerve VII (facial nerve, chorda tympani branch) and nerve IX (glossopharyngeal nerve, lingual branch). **ST** = solitary tract, **NST** = nucleus of the solitary tract, **Th.** = thalamus (arcuate nucleus), **Cx.** = cortex (taste area in postcentral gyrus). Associated pathways shown with dashed lines: **HN** = hypoglossal nucleus, **Amyg.** = amygdala.

bitter substances and is innervated by the glossopharyngeal nerve. Thus, a lesion to the glossopharyngeal nerve would reduce sensitivity to bitter taste with little effect on the other taste qualities.

The receptor potential of single receptor cells and the action potentials of single afferent fibers in response to chemical stimuli can be recorded. In general, each cell or fiber has its own pattern of response, responding maximally to one or more stimulus qualities and less to

others. Since a single fiber can respond to more than one stimulus quality, the unambiguous quality of, for example, saltiness probably requires the central nervous system to correlate the activity of a number of different fibers. There is a parallel in the auditory system: since each auditory nerve fiber responds to a range of frequencies, determining the pitch of a tone presumably requires the correlation of the activity of a number of fibers to determine which is responding the most.

In addition to quality, another characteristic of taste sensation is intensity. Judgments of stimulus intensity could be based on information provided by afferent fibers in the form of spike frequency, which is a function of the concentration of the stimulating chemical.

Behavioral Responses to Taste Stimuli

Human fetuses can swallow amniotic fluid as early as 12 to 16 weeks gestation. This response

Table 8–1. *Examples of Substances Eliciting the Four Taste Qualities*

Sweet	Sour	Salty	Bitter
Glucose	Citric acid	Sodium chloride	Quinine
Fructose	Acetic acid	Potassium chloride	Nicotine
Saccharin	Tartaric acid	Ammonium chloride	Caffeine
D-leucine	Hydrochloric acid	Magnesium chloride	Sodium citrate

is evidently influenced by the taste of the fluid, since with the addition of saccharin as part of a medical procedure, swallowing has been found to increase. Newborn infants demonstrate strong taste preferences, which have been studied by placing drops of distilled water or a weak solution of sugar, salt, or other solution on the tongue [4]. Sugar solution placed on the sweet-sensitive tip of the tongue causes an intake reflex: the tip of the tongue curls upward, and the tongue retracts in such a manner that liquid is moved toward the opening of the pharynx. The tongue movement is controlled by the hypoglossal nerve (figure 8–1). In addition, the infant may demonstrate associated facial expressions, such as a smile, that communicate a feeling of enjoyment. Once liquid reaches the opening of the pharynx, it can activate a swallowing trigger zone and thereby elicit a swallowing reflex.

Conversely, sour or bitter solutions lead to a rejection reflex, in which the tongue moves the solution out of the mouth.

In animals, behavior tests have been designed to measure the amount of preference or aversion elicited by various solutions [27]. Sucrose and sodium chloride elicited preference (drinking) in moderate concentration, but aversion in higher concentrations. The preferred concentrations could be altered by deprivation: salt-deprived animals preferred a more concentrated solution, an adaptive response in that it would help them maintain salt homeostasis.

Such experiments, and others in human subjects, have led to speculation that, under natural conditions, humans might naturally prefer the foods they need for proper nutrition. If so, such adaptive eating behavior could be threatened by processed and artificial foods. For example, a breast-fed infant will be encouraged to drink its mother's milk by the sweet taste of the lactose in the milk; at the same time the infant obtains the protein, fat, and other nutritional components of the milk. Later, the sweet taste of fructose will encourage the child to eat nutritious, ripe fruit. But if soft drinks containing sugar are taken with equal zest, no such nutritional benefits accompany them.

Smell

While the sense of taste responds to chemical substances already in the mouth, smell responds to chemical substances in the air, which may emanate from a distant source. Thus

Figure 8–2. *Distribution on the tongue of sensitivity to the four qualities of taste.*

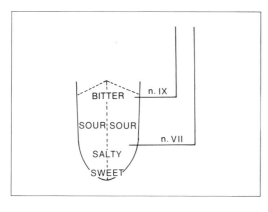

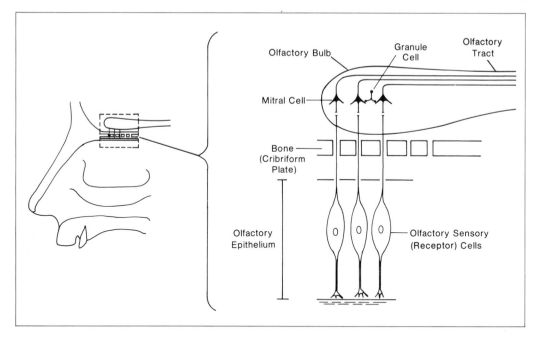

Figure 8–3. The olfactory epithelium, bulb, and tract. Left: Location with respect to the nasal cavity. Right: Neural connections.

appetizing foods may be approached and odoriferous poisons avoided while there is time to do so.

Structure

Olfaction is a scientific term for the sense of smell. The olfactory sensory cells are located in an area at the top of the nasal cavity (figure 8–3). The sensory cells together with supporting cells form an olfactory epithelium. The sensory cells have cilia projecting down into the mucous layer on the epithelial surface, and give off axons that pass upward through holes in the cribriform plate, a section of the ethmoid bone of the skull, into the olfactory bulb. There they synapse with axons of the large mitral cells, which are the principal cells in this synaptic region. The axons of the mitral cells form the olfactory tract, a nerve bundle on the underside of the frontal lobe, which sends projections to the olfactory bulb on the opposite side and to several regions of the brain. These regions include an olfactory area of the cortex, located on the ventral and medial surfaces of the frontal and temporal lobes, and the amygdala.

Odor Sensations and Sensory Responses

To elicit an odor sensation, a substance must be in gas or vapor form and then soluble in the mucous layer covering the olfactory epithelium. The qualities and classes of odor are more numerous and less agreed on than those for the sense of taste. The intensity of an odor increases with the concentration of the odorous substance. Odor sensation is notably subject to adaptation: a faint scent of perfume may disappear altogether after a few minutes of exposure.

A slow negative wave can be recorded from the olfactory epithelium, in response to an odorant stimulus, and reflects receptor potentials in the olfactory sensory cells. The wave increases with concentration of the stimulus. Like the taste fibers, individual olfactory-receptor cells are stimulated by several substances, more by some than by others. Action potentials recorded from individual nerve fibers in the olfactory tract either increase or decrease in frequency in response to specific odorant stimuli.

The synaptic organization of the olfactory bulb has received much experimental attention [31]. Interneurons called *granule cells* seem to lack axons; instead, their *dendrites* synapse upon the dendrites of the mitral cells. Another unusual feature is that the granule cells act by means of graded potentials rather than action potentials. Their effect is to inhibit the mitral cells; this can be compared to the effect of amacrine cells on ganglion cells in the retina.

Review Exercises

1. Describe the taste receptors and their neural connections.
2. List the four basic qualities of taste, and give examples of each.
3. Describe the behavioral responses to taste stimuli. Explain why they are generally adaptive.
4. Describe the olfactory receptors and their neural connections.
5. Describe the physiological responses that have been recorded from the olfactory system.
6. A patient has lost the sense of taste on the anterior part of the tongue; however, the ability to taste quinine and other bitter substances on the posterior third of the tongue remains. In what location is a lesion suspected?

9. Posture and Movement

How motor systems control the skeletal muscles

A tennis player waiting for the ball typically assumes a modified standing posture, erect but with knees slightly flexed and trunk bent forward at the waist. When the ball approaches, this posture is dissolved; the player runs to get into position and swings the racquet. Two elements of motor behavior are shown in this example: posture (the upright stance while waiting for the ball) and movement (the run and swing). Motor behavior can also be divided into reflex and supraspinal components. In the standing posture, for example, the muscle that extends the lower leg (the quadriceps femoris muscle, in the front part of the thigh) must contract to resist the pull of gravity. It can be shown that this contraction is largely due to a spinal reflex, an automatic response to gravity built into the neuronal circuitry of the spinal cord. But the strength of this reflex, and the changes that occur when the player moves toward the ball and then swings the racquet, are controlled by motor centers in the brain; they are subject to supraspinal motor control. Such movements are of course not built into every human being (or else everyone would be a good tennis player); but are matters of will, purpose, and learning. They are called voluntary, or purposeful, movements to distinguish them from purely automatic reflex responses.

Spinal Reflexes

A reflex is stereotyped motor response to a sensory stimulus. It is stereotyped in that, in general, all normal members of the same species respond in the same way to the same stimulus. Spinal reflexes are those responses whose essential neuronal pathways are contained within the peripheral nerves and spinal cord, although, as we shall see, they may be subject to descending influence from the brain. The simplest spinal reflex is the muscle stretch reflex; others are the inverse muscle stretch reflex and the flexor withdrawal reflex.

Muscle Stretch Reflex

An example of the muscle stretch reflex is the Achilles tendon or ankle jerk reflex, as shown in figure 9–1. Starting with the subject's leg relaxed, the stimulus (1) is a tap of the Achilles tendon provided by a reflex hammer. This causes an almost immediate stretch (2) of the gastrocnemius muscle in the calf. After a short delay, the reflex response is (3) a contraction of the gastrocnemius muscle, causing a plantar flexion of the foot.

The electrical activity associated with the muscle contraction can be viewed by placing recording electrodes (metal disks) on the skin overlying the muscle and amplifying the electrical potential difference between them. The electrical response recorded in this manner is called an *electromyogram* (EMG); it represents a compound potential due to the action potentials of numerous muscle fibers beneath the electrodes.

The EMG provides a measurement of the time interval, or *latent period*, between the

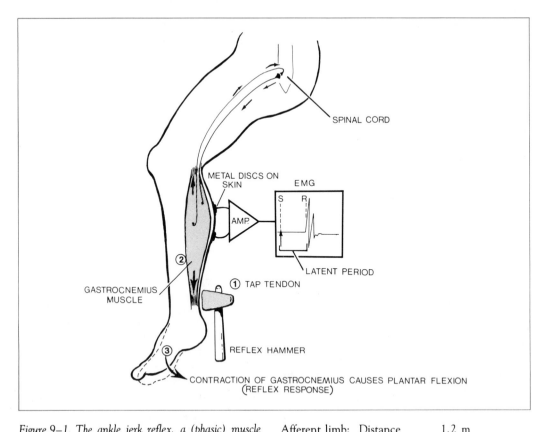

SPINAL CORD

METAL DISCS ON SKIN

EMG

S   R

AMP.

LATENT PERIOD

②

① TAP TENDON

GASTROCNEMIUS MUSCLE

③

REFLEX HAMMER

CONTRACTION OF GASTROCNEMIUS CAUSES PLANTAR FLEXION (REFLEX RESPONSE)

Figure 9–1. The ankle jerk reflex, a (phasic) muscle stretch reflex. **Amp.** *= amplifier,* **S** *= stimulus,* **R** *= response,* **EMG** *= electromyogram.*

stimulus and the response. In the example shown (in figure 9–1), the latent period is 30 msec. Most of the latent period can be attributed to conduction time along the afferent limb of the reflex (the nerve fibers from the muscle-stretch receptor to the spinal cord) and the efferent limb (the nerve fibers from the spinal cord to the muscle). The distance between the center of the muscle and the cord is approximately 1.2 m. The afferent nerve fibers are members of Group Ia, with velocities of approximately 100 m/sec. The efferent nerve fibers are A-alpha motoneurons, with velocities of approximately 80 m/sec. The resulting conduction times can be calculated as follows:

Afferent limb:	Distance	1.2 m
	Velocity	100 m/sec
	Conduction time	1.2 m/100 m/sec)
		= 12 msec
Efferent limb:	Distance	1.2 m
	Velocity	80 m/sec
	Conduction time	1.2 m/(80 m/sec)
		= 15 msec
Total conduction time		= 27 msec

The remaining 3 msec of the latent period is composed of additional delays due to excitation of muscle receptors, synaptic delay, and excitation of muscle fibers.

The basic phenomenon in the muscle stretch reflex is that a stretched muscle responds by actively contracting. Although in the ankle jerk reflex, the stretch is induced by tapping the tendon, more common stimuli are gravity or other forces during everyday life. Three examples are listed below.

1. During standing, the body tips forward under the influence of gravity, causing dorsal flexion of the foot relative to the lower leg. This stretches the gastrocnemius muscle. The resulting reflex contraction of the same muscle causes plantar flexion of the foot, assisting in the maintenance of an upright posture.
2. The influence of gravity during standing also causes the knees to bend, stretching the quadriceps femoris muscle in the anterior thigh. The resulting reflex contraction of the quadriceps femoris muscle causes the lower leg to straighten at the knee, assisting further in the maintenance of an upright posture.
3. When the hands are outstretched to catch a falling weight, the weight causes a brief extension of the arm at the elbow and, therefore, a stretch of the biceps muscle in

Figure 9–2. The muscle stretch reflex, showing the neuronal connections. **A** = *afferent limb of the reflex,* **B** = *efferent limb of the reflex,* **C** = *gamma efferent control,* **e** = *excitatory synapse.*

the upper arm. The resulting reflex contraction of the same muscle helps to keep the arm in its original position.

A common element in this reflex is that the muscle that is stretched is itself contracted. In addition, antagonist muscles at the same joint are automatically relaxed, so that the opposing muscle groups do not fight each other. This relaxation of antagonists is called *reciprocal inhibition.*

Neuronal Mechanism. In the muscle stretch reflex, the basic reflex arc is shown in figure 9–2, consisting of an afferent limb, A, and an efferent limb, B (ignore part C for the moment). On the afferent side, the stretch receptor consists of thin muscle fibers (intrafusal fibers) contained within a spindle-shaped capsule (the muscle spindle). The annulospiral (primary) sensory-nerve endings spiral around the thin muscle fibers in the expanded, fluid-filled region at the center of the capsule, and send impulses to the spinal cord via large-diameter, Group Ia sensory-nerve fibers, which conduct impulses at a high velocity (about 100 m/sec). The intrafusal muscle fibers lie parallel to the ordinary, extrafusal muscle fibers outside

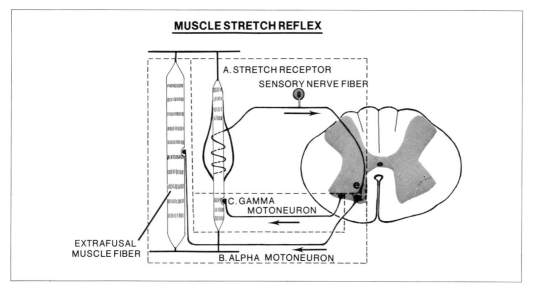

the spindle that make up the bulk of the muscle and cause it to contract.

The group Ia afferent fibers enter the cord via the dorsal root and pass through the gray matter to the alpha motoneurons in the ventral horn. There they end in excitatory synapses (labeled e in figure 9–2). Because only a single synapse in the CNS is involved, the muscle stretch reflex is monosynaptic and is the simplest reflex in the nervous system.

The efferent limb of the reflex labeled B in figure 9–2 consists of the alpha motoneurons, their axons, which exit the cord via the ventral roots and travel via the peripheral nerve to the muscle, and their neuromuscular junctions on the extrafusal muscle fibers. The result of nerve impulses traveling along this pathway is the contraction of the muscle.

The basic dynamics of the reflex are shown in figure 9–3. When the muscle is at its normal resting position, there is typically a fairly steady resting discharge from the stretch receptor and its sensory nerve fiber (figure 9–3A). When

the muscle is stretched, the spiral of nerve ending is pulled apart, causing an increased spike frequency as a response (figure 9–3B). The increased spike frequency lasts throughout the stimulus (the stretch receptor is a slowly adapting receptor). The afferent nerve impulses enter the cord and travel to excitatory synapses in the ventral horn. Alpha motoneurons are then activated and, in turn, activate the extrafusal muscle fibers, causing the reflex response—the muscle contracts and shortens (figure 9–3C). If there were no other inputs to the muscle spindle, this shortening would lead to slackness in the intrafusal fiber and an absence of response, or silent period, as shown in figure 9–3C.

Note that the reflex response is a contraction of the same muscle that was stretched. This in effect counteracts the stretch. At the same time, other muscles acting at the same joint, both synergists (which act in the same way as the stretched muscle) and antagonists (which act in the opposite way) are influenced by their motoneurons in a manner that supplements the reflex response.

We can examine some of these events at a synaptic level. In brief, the alpha motoneuron going to the stretched muscle is activated—it has an EPSP followed by a spike. At the neuromuscular junction, release of acetylcholine is followed by an end-plate potential, which leads to a muscle action potential and then muscle contraction. The alpha moto-

Figure 9–3. Sequence of events in the muscle stretch reflex. (A) Muscle at rest, showing resting discharge in Group Ia afferent fibers from stretch receptors. (B) Muscle is stretched (increased in length), showing increased afferent response. (C) Muscle contracts (the reflex response), showing the silent period in the afferent response that would occur without gamma efferent activity; normally, this silent pause would not occur because the gamma efferents would reduce the slack in the intrafusal muscle fibers.

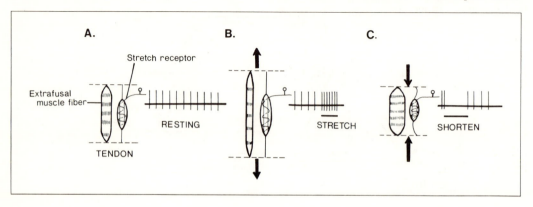

Table 9–1. Alpha and Gamma Efferent Fibers Compared

Type of Efferent Fiber	Typical Diameter	Typical Velocity	Destination
Alpha	14 μ	80 m/sec	Extrafusal muscle fibers
Gamma	4 μ	25 m/sec	Intrafusal muscle fibers

neurons going to the synergist muscle are facilitated—they have subthreshold EPSPs not followed by a spike but that make it easier for other imputs to cause spikes. The alpha motoneurons to antagonist muscles are inhibited—they have IPSPs, which block spike generation and leads to relaxation of antagonist muscles.

Gamma Motoneurons

Alongside the cell bodies of alpha motoneurons in the cord are smaller gamma motoneurons, which send gamma efferent fibers to innervate the thin, intrafusal muscle fibers within the muscle spindle (figure 9–2). The gamma efferent fibers are smaller in diameter and lower in conduction velocity than the alpha efferent fibers, as shown in table 9–1.

Gamma efferents end in neuromuscular junctions that are located only on the end (polar) regions of the intrafusal fibers. Their activation results in contraction of these end regions, resulting in a stretch of the central region with its associated sensory nerve endings. One effect of this gamma innervation is to take up the slack on the intrafusal fibers that would otherwise occur in a contracting muscle, as shown in figure 9–3C, and thus prevent the silent period in firing of the stretch receptor. Gamma activity, by filling in the silent period, eliminates this possible gap and helps maintain a continuous feedback to the CNS of information on muscle length, which is helpful in regulating muscle length. Another effect of gamma activity is to make the muscle stretch receptors more sensitive. This sensitivity control can be turned

up and down by fibers from the brain that innervate the gamma motoneurons.

Supraspinal Control

Both alpha and gamma motoneurons receive synaptic input from descending tracts originating in the brain (in supraspinal centers). Major descending tracts are the corticospinal tract from the cerebral cortex and the rubrospinal tract, vestibulospinal tract, and reticulospinal tract from the brain stem motor areas.

In brief, the supraspinal inputs act to:

1. *Control the muscle stretch reflex via a normal balance of excitatory and inhibitory synaptic inputs onto the motoneurons.* Lesions of supraspinal motor structures often may result in an imbalance that favors excitation. The muscle stretch reflexes then show a release of function and become hyperactive.

2. *Initiate voluntary activity.* The same spinal motoneurons involved in reflex activity can also participate in a voluntary movement when properly activated from brain centers. Sudden activation of either alpha or gamma motoneurons may result in muscle contraction. A *direct* pathway involves activation of alpha motoneurons, which then activate extrafusal muscle fibers. A possible *indirect* pathway involves activation of gamma motoneurons, which then activate the ends of the intrafusal muscle fibers, which in turn cause an afferent response in the stretch receptors and their afferent fibers, which in turn excite the alpha motoneurons and lead to a reflex contraction of the muscle as a whole. Clearly the indirect

pathway is more delayed. Actually, recent research shows that some human voluntary movements involve both direct and indirect pathways activated at about the same time, a phenomenon called *coactivation.*

Clinical Examination of Stretch Reflexes

Reflexes are examined by giving the patient a stimulus (input), and observing the response (output). Different time-courses of stimulus and response results in either phasic or tonic stretch reflexes. *Phasic stretch reflexes* result from a brief stimulus providing a sudden and transient stretch of the muscle. Tapping a tendon provides such a stimulus. Thus, the *tendon jerk* or *deep tendon* reflexes are phasic. The brief input results in a reflex contraction that is also brief. *Tonic stretch reflexes* result from a long-lasting stimulus, such as that provided by gravity. The long-lasting stimulus results in a long-lasting muscle contraction. During normal posture, this continuous muscle contraction is known as muscle *tone.* Tone is also reflected in a normal resistance to passive movement: when an examiner tries to dorsiflex a foot, there should be some resistance, provided by a tonic stretch reflex in the gastrocnemius muscle.

Tendon jerk reflexes may be elicited at a number of sites involving spinal roots at various segments of the cord. For example, the biceps jerk involves spinal segments C5 and C6, the triceps jerk involves C6 and C7, the knee jerk involves L2 to L4, and the ankle jerk involves S1.

It is important to compare the same reflexes on the right and left sides to see if they are symmetrical. Tendon jerk reflexes may be small or absent throughout the body in some cases, presumably when the excitability of alpha and gamma motoneurons is low. Reflexes can then be reinforced by asking the patient to voluntarily contract some other muscles, as in Jendrassik's maneuver where the hands are clasped together in front of the chest and pulled apart; this appears to increase motoneuron excitability in a general manner. On the other hand, in some patients the tendon jerk reflexes may be generally brisk, presumably due to an increased motoneuron excitability, for example in tense people.

The tonic stretch reflexes can also be examined clinically. Absent or weak stretch reflexes can result in reduced muscle tone, or flaccidity. Hyperactive stretch reflexes can result in increased muscle tone.

The absence of a muscle stretch reflex may indicate some lesion of either afferent or efferent limbs, or the cord at the appropriate segmental level. Such a lesion affecting spinal motoneurons may be called a lower motoneuron lesion. Conversely, hyperactive stretch reflexes may suggest a lesion in the descending fibers from supraspinal motor centers, which may be called an upper motoneuron lesion. For cerebral lesions such effects tend to be crossed; that is, a lesion affecting descending fibers from the right hemisphere will lead to hyperactive stretch reflexes on the left side of the body. A hyperactive stretch reflex may give rise to *spasticity,* a heightened resistance to passive movement, and in some cases to *clonus,* in which reflex contraction dies away and returns rhythmically, leading to a series of muscle contractions at a rate of about six per second.

Inverse Stretch Reflex (Golgi Tendon Organ Reflex)

The inverse stretch reflex was so named because it causes stretch of a muscle, at least past a certain point, to result in reflex relaxation of the same muscle, which is opposite to the action of the stretch reflex. Clinically, this is particularly evident in some spastic limbs: for example, passive flexion of a joint initially leads to great resistance, due to a hyperactive stretch reflex, but after a point, the resistance collapses, due to the inverse stretch reflex. In such cases, the inverse stretch reflex is thought to protect the tendon from being torn by excessive tension. In ordinary life, this reflex is thought to regulate the tension of a muscle, as opposed to the stretch reflex which regulates the length of a muscle. In many activities, such as, paddling a canoe, one might wish to maintain a fairly constant muscle tension even while joint angles and muscle lengths are continuously changing, and here the inverse stretch reflex could play a role.

Neuronal Mechanism. The receptors are the Golgi tendon organs (GTOs) located in the muscle tendon (figure 9–4), in series with the muscle fibers. These receptors respond to increased tension, whether due to an externally applied stretch of the muscle (passive stretch), or to contraction of the muscle itself, particularly against an external resistance, as in isometric contraction when the resistance pre-

vents overall change in muscle length. The GTO impulses travel to the cord along the large Group Ib afferent fibers, which are only slightly smaller than the Group Ia afferents from the annulospiral stretch receptors.

In the cord, the synaptic effect of these impulses from the GTOs is to exert inhibition (labeled *i* in figure 9–4) on the alpha motoneurons leading to the same muscle. There is probably a short, inhibitory interneuron in the reflex arc between the afferent and efferent neurons; thus, the reflex is thought to be disynaptic. The effect of inhibiting the impulses in the alpha motoneuron is to cause relaxation of the muscle, which is the reflex response.

Flexor Withdrawal Reflex

The flexor withdrawal reflex is typically the response of an arm or leg to a painful or hot stimulus, which results in the withdrawal of the limb from the stimulus, associated with flexion at several joints in the limb. Common exam-

Figure 9–4. The inverse stretch reflex; **e** = *excitatory synapse,* **i** = *inhibitory synapse.*

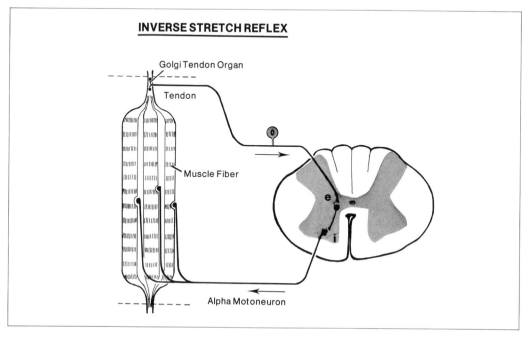

ples are when you pull your hand away from a hot frying pan or lift your foot away from a thumbtack. In these examples, the reflex response generally begins before you are consciously aware of the painful stimulus—because the spinal reflex arc can function before the afferent impulses have affected the brain. In the second example, the opposite foot must support the body alone; it is aided in this task by an associated reflex called *crossed extension,* which activates motoneurons to the extensor muscles of the opposite leg to withstand the pull of gravity.

Neuronal Mechanism. The receptors are pain and temperature receptors in the skin, giving rise to Group III (or A-delta) and Group IV (or C) fibers which enter the cord via the dorsal roots. In chapter 5, we saw how these fibers synapse in the dorsal gray horn of the cord and relay pain and temperature information to neurons whose axons cross over to ascend via the lateral spinothalamic tract to the brain. In addition (figure 9–5), these dorsal root fibers

Figure 9–5. The flexor withdrawal reflex. **S.G.** = *substantia gelatinosa,* **e** = *excitatory synapse.*

also excite a chain of interneurons in the gray matter of the cord that eventually excite a number of alpha motoneurons in the ventral horn of the cord. The motoneurons are diffuse in that they may be located in several segments of the cord, and their effect is to cause the contraction of flexor muscle fibers in the several joints of the affected limb. The spread of activity from a limited input to a widespread output is called *divergence.*

The reflex is termed *polysynaptic* because a number of CNS synapses are involved. The greater latent period, as compared with that of the muscle stretch reflex, is attributed in part to the large number of synaptic delays in the reflex arc. The reflex response may continue for several seconds after a strong stimulus is stopped; this afterdischarge may be due to the interneuronal circuits, some of which are circular and lead to repeated excitation of the same pathway.

Other Reflexes and Responses

In addition to the elementary spinal reflexes, a number of other reflexes and automatic pos-

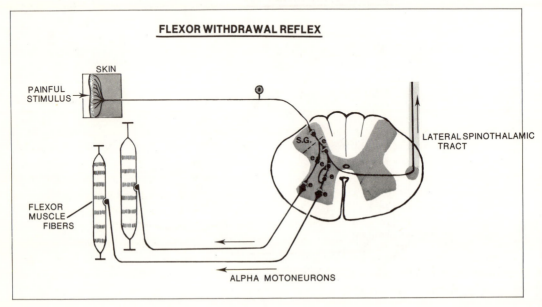

tural responses have been described. Several of these, such as the tonic neck reflex, can be seen in infancy, disappearing with the maturation of the supraspinal motor mechanisms. Only three examples will be listed here. Although largely spinal, they are subject to supraspinal influence.

Tonic Neck Reflexes. When the head is turned to the right, the right arm and leg are extended, and the left limbs are flexed; when the head is turned to the left, the left limbs are extended, and the right limbs are flexed. This resembles, and might have led to, the pose of the boxer or archer. The sensory receptors are in the joints of the cervical vertebrae.

Positive Supporting Reaction. When the sole of the foot touches the ground, both the extensors and flexors of the leg contract, making the leg a rigid pillar, as in the standing posture.

Plantar Reflexes. The stimulus is applied to the outer border of the sole of the foot by scratching it with a key or other blunt object. The normal response, after the first year of life, is a plantar flexion (downward bending) of the toes, usually combined with upward flexion of the foot at the ankle. During the first year of life, before the supraspinal motor systems (in particular the pyramidal or corticospinal tract) have completely developed, the response is an upward movement of the big toe. In adults, a similar upgoing toe response can be found during sleep, and also with lesions of the pyramidal or corticospinal tract. Another term for the upgoing toe response is *Babinski's sign.*

Supraspinal Motor Control

The spinal reflex mechanisms discussed in the previous section are normally under the control of supraspinal motor centers (higher centers, motor areas in the brain). These centers are of major importance physiologically and clinically. Unfortunately for the student, the structure and activity of these centers are complex and not completely understood. The approach of this book is to simplify this subject; more advanced textbooks may then be consulted for further detail.

The four major centers to be described here are the motor cortex, the brain stem motor centers, the basal ganglia, and the cerebellum. These areas interact with each other, and with spinal reflex mechanisms including afferent fibers and motoneurons.

Motor Cortex

Structure. The motor cortex is located in the frontal lobe of the cerebrum; it includes the precentral gyrus (just anterior to the central sulcus), and is also known as Brodmann's area 4.

The motor cortex is a layered structure featuring [31] large cells in layer V that because of their shape are called *giant pyramidal cells* and send axons to the spinal cord; the length of some of these axons exceeds one meter. The axons of these and other pyramidal cells form the *corticospinal tract,* so named because it is a direct pathway from the cortex to the spinal cord (figure 9–6). In the cerebrum, as this tract passes between the thalamus and the basal ganglia, it is part of the internal capsule, a structure of great clinical importance as a frequent site of damage in cerebrovascular accidents (or strokes); in the medulla, it forms two prominent ridges along the ventral surface known as the medullary pyramids. In the medulla, near its border with the spinal cord, these fibers cross over to the opposite side and then descend the cord as the lateral corticospinal tract. At various segments of the cord, fibers turn off from this tract to synapse on motoneurons in the ventral horn, either di-

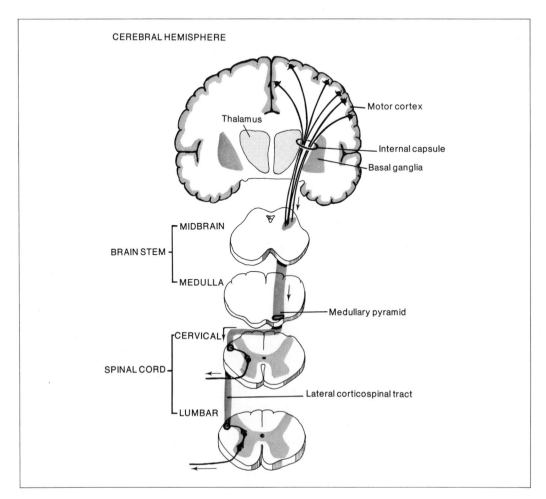

Figure 9–6. The lateral corticospinal tract, from neurons in the motor cortex to spinal motoneurons.

rectly or indirectly via interneurons. A smaller number of fibers (about 10%) do not cross and form the *medial corticospinal tract*. Other fibers from motor cortex project to brainstem motor areas, basal ganglia, and cerebellum.

Somatotopic Mapping. Artificial stimulation of area 4 with low-voltage electrical pulses can cause movement of various parts of the body. The pulses activate pyramidal cells in a particular part of the motor cortex, and the resulting movements show which parts of the body the

axons of these cells control. These movements are mostly on the contralateral (opposite) side of the body, by virtue of the largely crossed nature of the descending tracts. When the induced movements are recorded on a diagram of the motor cortex, the result is a somatotopic map—a map of the body on the motor cortex.[1] On this map the legs are located on the medial surface of the cerebral hemisphere, and the hands and face are located ventrolaterally, closer to the temporal lobe. The map is distorted in that a disproportionately large area

[1]Another type of somatotopic map is found on the somatic sensory area of the cortex, as described in chapter 5.

of the cortex is devoted to control of the hands and face, areas capable of the complicated and precise movements.

Most neurons that originate in the motor cortex and project via the corticospinal tracts affect spinal motoneurons indirectly via one or more spinal interneurons. However, some neurons project directly to spinal motoneurons, particularly those that govern discrete movements of finger muscles. Both cases are illustrated in figure 9–6.

Function. The motor cortex functions to initiate voluntary movement and to control spinal reflexes. Simple voluntary hand movements, such as pressing a button with the thumb, have been studied in humans and animals [8,33] enough to present a tentative outline of the neural events involved. About 100 msec before the movement, some motor cortex neurons begin firing; others, which were tonically active before, now stop firing. After a time interval necessary for conduction down the corticospinal tract and synaptic delay, spinal motoneurons in the cervical cord are activated. Both alpha and gamma motoneurons appear to be activated at about the same time, a process known as coactivation. The alpha motoneurons cause muscle contraction directly, while the gamma motoneurons work indirectly via the muscle stretch reflex arc.

Destructive lesions of motor cortex (area 4) or of its descending projections in the internal capsule or other levels of the brain have the following effects:

1. Voluntary movements are diminished (paresis) or lost (paralysis) in the corresponding parts of the body, on the side opposite to the lesion.
2. Certain spinal reflexes are altered. In particular, a positive Babinski's sign appears, in which stroking of the sole of the foot results in upward extension of the big toe. After infancy this reflex is normally suppressed by

corticospinal fibers. Other changes, such as enhanced stretch reflexes and spasticity, are due to cortical lesions that extend beyond area 4 to area 6 and involve a relay in brainstem motor areas. Spinal-reflex changes are also on the side opposite to the brain lesion.

Brainstem Motor Areas

In addition to the direct pathway from motor cortex via internal capsule, pyramidal tract, and corticospinal tract to spinal motoneurons, there are important indirect pathways in which the motor cortex relays information via several brainstem motor areas and their descending tracts to spinal motoneurons. These brainstem areas also receive inputs from noncortical sources. The brainstem motor areas and their descending tracts contribute to the normal control of posture and movement and also account for many signs of motor disorder after brain lesions.

Structure. Three brainstem motor areas are described here (figure 9–7).

1. The *red nucleus* (*nucleus ruber*) is located in the midbrain, the highest division of the brainstem. It receives inputs from the motor cortex, cerebellum and possibly the basal ganglia. It gives rise to the rubrospinal tract, which descends in the lateral column of spinal white matter, adjacent to the lateral corticospinal tract.
2. The *reticular formation* is a column of neurons at the core of the brainstem ventral to the fourth ventricle. The many cells with short, branching axons and their numerous synapses running the length of the brainstem have an appearance that is netlike or reticulated, giving rise to the name. (See more detailed texts for discussion of lateral, medial, and other subdivisions.) The reticular formation receives inputs from the motor

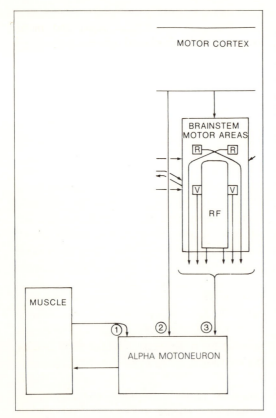

Figure 9–7. Brainstem motor areas and the descending tracts that carry their output. R = *red nucleus,* V = *vestibular nuclei,* RF = *reticular formation; their descending tracts are, respectively, rubrospinal, vestibulospinal, and reticulospinal tracts, and may be grouped together as the extrapyramidal tracts or extracorticospinal tracts. These tracts, labeled* 3, *influence alpha motoneuron along with corticospinal tract,* 2, *and muscle stretch afferents,* 1.

cortex, cerebellum, and basal ganglia, among other sources. It gives rise to the reticulospinal tracts, which descend in the spinal white matter.

3. The *vestibular nuclei,* located in the pons, are sensory nuclei of the vestibular nerve that also have a motor function. Their inputs are from the vestibular organs and the cerebellum, and they give rise to the vestibulospinal tracts, which descend in the ventral and medial portions of the spinal white matter.

These three descending pathways are often called *extrapyramidal pathways* because they travel outside the pyramidal tracts. In the spinal cord, the rubrospinal tract travels alongside the lateral corticospinal tract, while the reticulospinal and vestibulospinal tracts are located more ventral and medial.

Function. The major function of these pathways is to gather information from other sources, including motor cortex, cerebellum, basal ganglia, and vestibular nerve, and use it in the control of spinal motoneurons.

The effects of the descending fibers in the extrapyramidal tracts are largely exerted on motoneurons indirectly through spinal interneurons. Both alpha and gamma motoneurons are affected, and both excitatory and inhibitory synaptic responses (EPSPs and IPSPs) have been shown.

The best-understood effect is on the upright posture. Standing upright involves contraction of the antigravity muscles, including those that extend the leg and hold the head up, and, in humans, those that flex the arm (in four-legged animals the forelimb extensors are antigravity muscles). The stretch reflexes of these muscles are enhanced by activation of a large part of the reticular formation and of the vestibular nuclei, at least in animals. Many of the descending fibers at the level of the internal capsule or midbrain, originating in the motor cortex and basal ganglia, normally function to keep the brainstem motor areas under control. The loss of these fibers after internal capsule or midbrain lesions is followed by a release of function in which the facilitating effects of brainstem motor areas are unchecked, leading to an exaggeration of the standing posture, hyperactive tendon jerk reflexes in the antigravity muscles, and spasticity. The particular posture that results varies with the level of the lesion. At the internal capsule level, the arms tend to be flexed. At the midbrain level, the arms tend to be extended. In both cases, the legs tend to be extended.

In addition to posture, the reticulospinal and vestibulospinal tracts help control the motoneurons innervating muscles of the central and proximal parts of the body, including the neck, back, shoulder, and hip joints, which are important in large, integrated movements of the body and limbs, such as turning, bending, or walking. The rubrospinal tract, in contrast, may assist the adjacent lateral corticospinal tract in controlling independent movements of the more distal muscles, such as those of the fingers [28].

Basal Ganglia

Structure. The basal ganglia are a group of nuclei located well beneath the surface of the cerebral hemispheres, separated from the thalamus by the fibers of the internal capsule. The major nuclei are the *caudate, putamen,* and *globus pallidus* (figure 9–8). The first two

Figure 9–8. Basal ganglia, indicating major neuronal connections between nuclei in the basal ganglia and relationships with other structures. **C** = *caudate nucleus,* **P** = *putamen,* **Sm** = *striatum,* **GP** = *globus pallidus,* **SN** = *substantia nigra,* **STN** = *subthalamic nucleus. Thalamic relay nucleus between basal ganglia and motor cortex is not shown.*

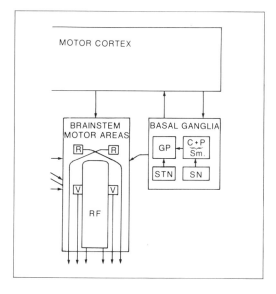

(together known as the *striatum*) receive afferent input from the association and motor cortex, and the third sends efferent output back up to the motor cortex (via a relay in the thalamus) and down to the brainstem motor areas. These nuclei connect to each other and also to other nearby nuclei that are often included in the basal ganglia: the *substantia nigra* and the *subthalamic nucleus.*

Compared with other brain regions, the basal ganglia have extremely high concentrations of two neurotransmitters, *acetylcholine* and *dopamine.* Neurons originating in the substantia nigra secrete dopamine at their synaptic terminals in the striatum, and acetylcholine is secreted at synapses between striatal neurons, according to present evidence [31]. These transmitters are of practical importance in diseases affecting the basal ganglia.

Normal and Disordered Function. Most of the interest in basal ganglia function derives from the movement disorders associated with lesions and dysfunction in the human basal ganglia. *Parkinsonism,* the most common movement disorder, has both negative and positive symptoms. The negative symptom (reflecting insufficient activity) is both a difficulty in starting and a decrease in voluntary movements (hypokinesis). The positive symptom (reflecting excessive activity) is an increased tone of both extensor and flexor muscles, leading to rigidity; the rigidity switches on and off at 5 to 10 c/s during passive movement of a limb (cogwheel rigidity) or at rest (resting tremor).

In most cases, examination of the brain by pathologists shows degeneration in the substantia nigra, associated with a reduction in the amount of dopamine in the substantia nigra and striatum. Normally, a group of nerve cells in the substantia nigra sends axons to the striatum, where they secrete dopamine; the axons are called nigrostriate dopaminergic fibers. A loss of these neurons is thought to be the underlying lesion in parkinsonism (figure 9–9A and 9–9B).

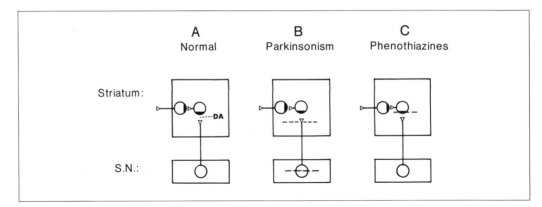

Figure 9–9. Proposed neuronal mechanism in parkinsonism. **S.N.** = substantia nigra, **DA** = dopamine. Dashed lines show sites of degeneration or chemical blockade. (A) Normal mechanism. (B) In parkinsonism, cells in substantia nigra and their dopamine-secreting terminals degenerate. (C) Phenothiazine drugs can block dopamine receptors, leading to side effects that resemble parkinsonism.

The presumed dopamine deficiency cannot be treated by administering dopamine itself, because it cannot cross the barrier from the circulation into the brain (the blood–brain barrier). The deficiency can be relieved by administering L-dopa (L-dihydroxyphenylalanine), which can cross the barrier and then be converted into dopamine. L-dopa provides symptomatic relief in many cases and represents a major advance in neurological treatment, in that it represents a partial replacement of a transmitter known from other evidence to be deficient.

A syndrome resembling parkinsonism can also be induced as a side effect of the phenothiazine drugs, such as chlorpromazine (Thorazine) and trifluoperazine (Stelazine), that are commonly prescribed in psychiatric practice. A possible basis for this side effect is that the phenothiazines block dopamine receptor sites in the postsynaptic membrane of neurons (figure 9–9C). Thus, it seems that similar symptoms can result from (1) degeneration of dopamine-secreting presynaptic terminals at synapses in the striatum (figure 9–9B), or (2) blockade of postsynaptic dopamine receptors at the same synapses (figure 9–9C). (The favorable effect of phenothiazines on psychiatric symptoms may be due to their action at other synapses.)

In addition to L-dopa, anticholinergic drugs such as atropine may also be given to treat parkinsonism. Their effect seems to be on cholinergic synapses in the striatum, whose action opposes that of the dopaminergic synapses. Drugs that mimic or increase the effects of acetylcholine tend to worsen parkinsonian symptoms.

Other movement disorders—chorea, athetosis, and ballismus—have mainly positive symptoms: involuntary movements, that is, movements over which the patient seems to have no control. In chorea, the movements are jerky or birdlike. In athetosis, the movements are writhing or snakelike. Both chorea and athetosis are associated with lesions in the striatum. In ballismus, the movements resemble those involved in flinging a ball. Ballismus is associated with lesions in the subthalamic nucleus, which is closely connected with the globus pallidus.

A possible interpretation for these movement disorders is that a controlling influence, possibly inhibitory, normally exerted by basal ganglia neurons on the motor cortex, is disturbed by such lesions, leading to excessive activity in the motor cortex. This interpretation is supported by some animal experiments on basal ganglia function. When electrical stimulation of the motor cortex was used to initiate a movement, the movement could often be stopped by stimulation of the striatum. In other experiments, spike activity recorded from motor cortex neurons was reduced in frequency by electrical stimulation of the striatum. Both effects could result from striatal inhibition of motor-cortex neurons. A corollary to this interpretation is that lesions in the substantia nigra increase the inhibitory activity of the same striatal neurons, leading to the negative symptoms of parkinsonism (reduced and slowed movement).

Other experiments showed that striatal neurons, recorded with microelectrodes, increased their firing rate at the beginning and end of voluntary movements, particularly when the movements were slow. This has led to speculation that the basal ganglia are involved in the programming (initiation, guidance) of slow voluntary movements.

The two different effects that striatal neurons may exert on motor cortex—inhibition and initiation of activity—are not necessarily in conflict. Initiating a new movement often requires an inhibition of the preexisting movement or posture (e.g., sitting down requires a suppression of the muscle contractions involved in the standing posture). Both effects could therefore be part of the same overall basal ganglia–motor cortex system.

Cerebellum

Like the basal ganglia, the cerebellum forms a closed loop of neural connections with the cerebral cortex and modifies cerebral output to spinal motoneurons.

Structure. The cerebellum is a large, roundish structure protruding from the dorsal surface of the brainstem at the pontine level. Like the cerebrum, it has a cellular cortex surrounding a white region of nerve fibers that contains several subcortical nuclei. All cerebellar output signals are carried by axons of the large *Purkinje cells,* the principal cells of the cerebellar cortex. These cells have many-branched dendritic trees, containing thousands of synapses from input fibers and the numerous interneurons (basket cells, granule cells, and stellate cells) that modify the activity of the Purkinje cell.

The three divisions of the cerebellum are shown in figure 9–10. The archicerebellum (Arch. in the figure), the most primitive part, has input and output connections with the vestibular nuclei, one of the brainstem motor areas described earlier in the chapter. The paleocerebellum (Pal.) receives afferent input from muscle spindles, Golgi tendon organs, joint receptors, and other peripheral receptors, via the rapidly conducting spinocerebellar tracts (and others), and sends output to the brainstem motor areas. The neocerebellum (Neo.), the major part in humans, receives input from several areas of cerebral cortex including the motor cortex, relayed through the pons; it sends output fibers back to the motor cortex, relayed through the thalamus.

Function. The archicerebellum developed in connection with the vestibular nuclei and functions with it in maintaining balance. Lesions of this area lead to nystagmus, head rotation, and/or difficulty in maintaining balance while walking.

The paleocerebellum and neocerebellum function in the control of posture and movement. One effect of cerebellar lesions, related to posture, is muscular hypotonus due to underactive muscle stretch reflexes. Other effects, related to voluntary movement, include the following:

1. *Dysmetria*—a tendency to overshoot or undershoot a target, for example, knocking

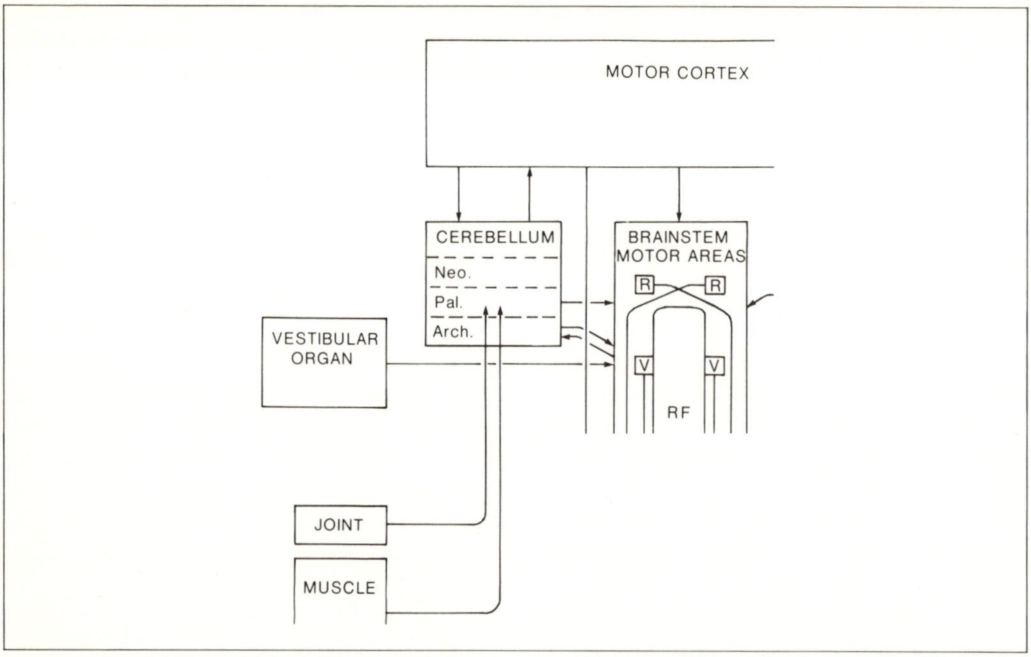

Figure 9–10. Cerebellum, indicating major divisions and relationships with other structures. **Arch.** = *archicerebellum,* **Pal.** = *paleocerebellum,* **Neo.** = *neocerebellum. Afferent information from joint and muscle receptors ascends in the spinocerebellar and other tracts.*

over a glass when reaching for it; failure to terminate a motion properly.

2. *Adiadochokinesis*—difficulty in making rapidly alternating movements, for example, inability to rotate the palm up and down quickly.

3. *Scanning speech*—difficulty in making the rapid movements of the vocal apparatus necessary for smooth speech.

4. *Intention tremor*—a tremor that appears during a voluntary movement rather than during rest.

These effects on movement have been related to the role of the cerebellum as a system that uses feedback signals to correct errors.

To understand the concept of error correction, you might watch a subject, with eyes closed and arms stretched out to both sides, who then attempts to bring together the middle fingers of both hands. Without visual information this action is difficult but still possible. Observe how errors (deviations from the target) are corrected, and how the movement slows down when the target is nearly reached. In everyday life, similar corrections in skilled movements, such as reaching for and grasping a cup, are done automatically and unconsciously, in large part by the cerebellum and its interactions with other structures. According to one model, even before the motor cortex initiates the movement, the cerebellum may be involved in planning (programming, sequencing) the movement. For this stage, the cerebellum receives input information from association cortex, processes it, and sends output information to motor cortex. At the initiation of the movement, the motor cortex sends signals not only to the spinal motoneurons but also to the cerebellum, informing it of the intended movement ("reach for the cup"). The signals sent to the cerebellum are processed within its complex neuronal circuitry and combined with feedback signals from muscle, joints, and other peripheral receptors, which provide information on the actual moment-to-moment position and speed of the hand as it approaches the cup. Output signals then are sent from the cerebellum back to the motor cortex and lead to correction of errors (for example, "more to the right," "slow down," "stop") during the movement.

On the basis of this model, cerebellar lesions could lead to deficiencies in planning voluntary movements, in correcting errors during the movements, or both. If, for example, the hand strays to one side of the target and the error is overcompensated by moving it too far to the other side, a back-and-forth movement (intention tremor) can result. If the hand moves too rapidly toward the target, and this error is not corrected in time, the hand will overreach its target (dysmetria).

An Overview of Motor Mechanisms

A schematic diagram of the various motor systems combined is shown in figure 9–11. Note the central position of the alpha motoneuron, receiving and integrating information from the following sources:

1   muscle stretch receptors (and other afferent input not shown).
2   the motor cortex, via corticospinal (pyramidal) tract fibers.
3   the brainstem motor areas, via the descending rubrospinal, vestibulospinal, and reticulospinal tracts (extrapyramidal tracts).

The result of all this information is muscle contraction determined by the frequency of action potentials coming from the alpha motoneuron. The motoneuron has therefore been called the *final common path* of the motor systems.

The tone of the antigravity muscles during the upright posture is maintained by the stretch reflex, utilizing afferent path **1** as shown in the diagram, with the strength of that reflex determined largely by tract **3**. Voluntary, purposive movement is associated with control signals in descending tracts **2** and **3**; the precise movements of the hands and fingers are particularly dependent on tract **2**.

Information descending through tracts **2** and **3** is produced by neural activity in four interrelated supraspinal motor systems: motor cortex, brainstem motor areas, basal ganglia, and cerebellum. The loop from cortex to basal ganglia and back suggests that motor cortex activity may be slowed down, inhibited, or otherwise modified by the basal ganglia. The loop from cortex to cerebellum and back suggests that the cerebellum could play a similar role. In addition, afferent input to the cerebellum from muscle, joint, and other sensory receptors, and the outcome of cerebellar lesions, suggests that the cerebellum could function as the central component of a servomechanism. That is, it could combine information from the motor cortex on the intended action (e.g., "reach for the cup") with information from sensory receptors on the current situation ("the hand is moving too quickly") to produce an error-correction signal ("slow down"). The error-correction signal could go back up to the motor cortex and/or be transmitted downward via the brainstem motor areas.

What precedes activity of the motor cortex in the execution of a voluntary movement? Neural activity in the following areas has been suggested: (1) other areas of cerebral cortex, especially the premotor cortex, Brodmann's area 6, and the prefrontal lobes, anterior to area 6; (2) basal ganglia; and (3) cerebellum. These possibilities are under investigation. For one important type of voluntary activity, speech, the influence of a frontal lobe region (known as Broca's area) anterior to the part of the motor cortex that controls the muscles of

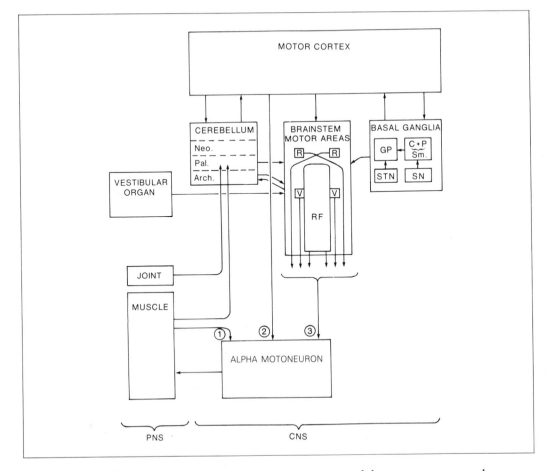

Figure 9–11. A schematic diagram showing the overall plan of the motor systems, combining figures 9–7, 9–8, and 9–10 (same abbreviations). Three major influences on alpha motoneuron activity are shown: (1) Muscle stretch afferents (Group Ia fibers), from muscle stretch receptors. (2) Corticospinal fibers, from motor cortex. (3) Descending fibers in the rubrospinal, vestibulospinal, and reticulospinal tracts (extrapyramidal or extracorticospinal tracts), from brainstem motor areas.

speech is known to be significant in organizing speech output.

Review Exercises

1. Diagram and describe the muscle stretch reflex, including the role of muscle spindle receptors, alpha motoneurons, and gamma motoneurons.

2. Give two examples of how the muscle stretch reflex helps to maintain an upright posture.

3. Describe the sequence of neuronal events in the muscle stretch reflex.

4. Describe the effects of activity in the gamma motoneurons.

5. Compare the activation of alpha motoneurons either directly or indirectly via the gamma motoneurons. Define coactivation.

6. Define phasic and tonic stretch reflexes. Describe how each can be tested clinically, giving examples.

7. Define muscle tone, flaccidity, spasticity,

and clonus. Relate each term to spinal reflexes.

8. Diagram and describe the inverse muscle stretch reflex.

9. Diagram and describe the flexor withdrawal reflex.

10. For the three major spinal reflexes described (muscle stretch, inverse stretch, and flexor withdrawal), list the following information in a table: effective stimulus, receptor, afferent fiber type (I, II, etc.), number of synapses in the CNS, and effect on motoneurons (showing which are excited and which are inhibited).

11. Define reciprocal inhibition as applied to spinal reflexes.

12. Describe the tonic neck reflexes, positive supporting reaction, and plantar reflex. Define Babinski's sign and describe its significance.

13. List in outline form the major supraspinal motor-control centers and their components and output tracts.

14. Describe the somatotopic map on the motor cortex and the effects of motor-cortex stimulation.

15. List the sequence of events in a voluntary movement.

16. Describe the functions of the brainstem motor areas, according to current information.

17. Describe the symptoms of parkinsonism and their relationship to dopamine deficiency, L-dopa medication, and undesirable side effects of the phenothiazine drugs.

18. Define the movement disorders of chorea, athetosis, and ballismus.

19. Diagram and describe the function of the cerebellum as a servomechanism utilizing feedback to control voluntary movement.

20. List the possible manifestations of lesions in the following areas: motor cortex, internal capsule, striatum, substantia nigra, archicerebellum, paleocerebellum, and neocerebellum.

21. Describe how alpha motoneurons integrate information from several sources.

10. Autonomic Function

How sympathetic and parasympathetic activity help control the internal environment

A novice skier about to start down a mountain slope is excited and a little anxious. His heart beats rapidly and with more force than usual, and his digestive processes seem to be shut down. The circulation to his skin is reduced, making it pale and cool. Sugar is released into his blood from storage sites within liver and muscle. His pupils are dilated.

All these various responses are due to activation of the sympathetic nervous system, one of the two divisions of the autonomic nervous system. The responses are also related in that they all can contribute to physical performance under challenging conditions, especially in the presence of danger or cold. The increased cardiac output, measured as liters of blood pumped per minute, is largely available to supply the skeletal muscles, since the requirements of the digestive organs and the skin are reduced. The muscles, about to produce extra work, will be nourished by the additional sugar and oxygen supplied by the blood. The dilated pupils allow obstacles to be seen even in dim light.

In contrast, hours later, the same person is relaxed and drowsy after a good dinner and a glass of beer. His heart rate is slow; digestion is active; skin is flushed and warm, supplied by an increased blood flow; and pupils are constricted. These responses are all due to activity in the parasympathetic nervous system, the other division of the autonomic nervous system. In many but not all instances, parasympathetic effects are associated with relaxation and rebuilding of stores of metabolic energy.

While the posture and movement of the body in the outside environment are controlled by the somatic motor nerve fibers, the function of the heart, blood pressure and flow, body temperature, and other variables of the internal environment are controlled in large part by the visceral motor nerve fibers of the autonomic nervous system. While the somatic motor nerve fibers control the skeletal muscles, the visceral motor nerve fibers control the smooth muscles, cardiac muscle, and exocrine glands. In both cases, motor output is influenced both by sensory input and by control centers in the brain.

Functional Anatomy

Sympathetic Nervous System

The sympathetic efferent nerves originate in a group of nerve cells running along the lateral horn of the spinal gray matter in the thoracic and upper-lumbar segments (from T1 to L3) of the spinal cord (figure 10–1). These cells send axons through the ventral roots to the chain of sympathetic ganglia alongside the cord or to more peripheral ganglia and are known as the *preganglionic sympathetic neurons*. Within the ganglia, preganglionic nerve fibers synapse onto *postganglionic sympathetic neurons*. Some preganglionic fibers travel several segments up or down the chain of ganglia before synapsing, and some synapse only on more peripheral ganglia. The postganglionic sympathetic axons

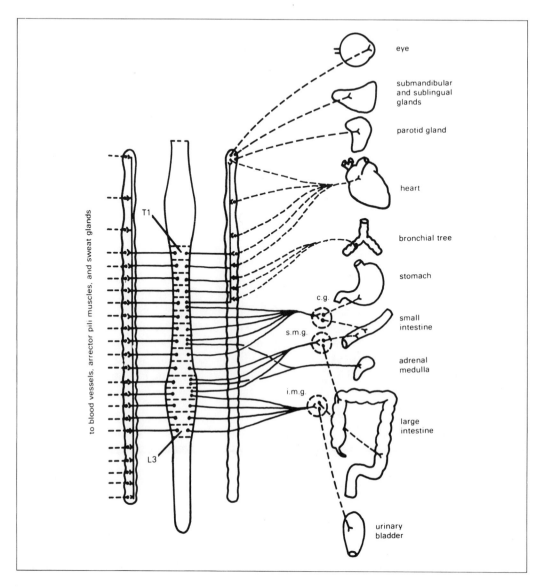

Figure 10–1. The sympathetic nervous system, showing efferent fibers; chains of sympathetic ganglia on both sides of the spinal cord; peripheral ganglia (c.g. = celiac ganglion, s.m.g. = superior mesenteric ganglion, and i.m.g. = inferior mesenteric ganglion); and effector organs. Source: Barr, M.L. The Human Nervous System: An Anatomic Viewpoint (3rd ed.). Hagerstown, Md.: Harper & Row Publishers, Inc., 1979. Reprinted with permission.

are slowly conducting, unmyelinated C fibers. They travel outward via the peripheral nerves to various effector organs, where they supply smooth muscle, cardiac muscle, and exocrine glands.

The secretory cells of the *adrenal medulla* have an embryonic origin similar to that of the postganglionic sympathetic neurons. Like those

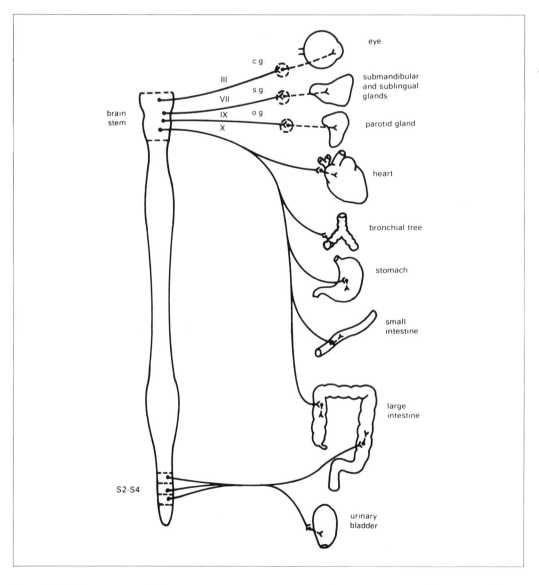

Figure 10–2. The parasympathetic nervous system, showing efferent fibers in cranial nerves III, VII, IX, and X, and in pelvic nerves; ganglia (c.g. = ciliary ganglion, s.g. = submandibular ganglion, and o.g. = otic ganglion); and effector organs. Source: Barr, M.L. The Human Nervous System: An Anatomic Viewpoint (3rd ed.). Hagerstown, Md.: Harper & Row Publishers, Inc., 1979. Reprinted with permission.

neurons, they are innervated by preganglionic sympathetic fibers. The secretions of the medullary cells, however, are released into the general circulation.

Parasympathetic Nervous System

Like the sympathetic system, the parasympathetic nervous system has preganglionic and postganglionic neurons (figure 10–2). The *preganglionic* parasympathetic neurons originate in the brainstem, sending axons outward via the oculomotor, vagus, and other cranial nerves, and in sacral segments of the spinal cord, sending axons outward via the pelvic nerve (also called *nervi erigentes*). The preganglionic nerve fibers are long, and synapse in parasympathetic ganglia, onto *postganglionic parasympathetic neurons*. Since the parasympathetic ganglia are located near or within their effector organs, the postganglionic axons are shorter than those of the sympathetic system.

Autonomic Reflexes

The activity of autonomic neurons is controlled in part by reflex activity at a brainstem or spinal level. Some examples of autonomic reflexes are given below.

1. Within the carotid and other large arteries, stretch receptors called *baroreceptors* are activated by increased arterial blood pressure, and transmit information to centers in the brainstem. Reflex output to the heart and the blood vessels generally causes the blood pressure to return toward a normal level. This is known as the baroreceptor reflex.
2. When stretch receptors within the walls of the bladder are activated by a full bladder, they transmit information to the sacral spinal cord. Within the cord, signals are passed via interneurons, leading to reflex output to the smooth muscles of the bladder. The bladder contracts and empties. A similar spinal reflex leads to emptying of the bowel.
3. Warming the skin can lead to increased blood flow and reddening of the affected area, while cooling the skin can have an opposite effect. These effects involve a spinal reflex originating in cutaneous thermoreceptors, with reflex output to cutaneous blood vessels.

Supraspinal Control of the Autonomic Nervous System

Like the muscle stretch reflex and other spinal reflexes that control posture and movement, the autonomic reflexes of the spinal cord also are normally under the control of the supraspinal centers. Thus, in the adult, emptying of the bladder and bowel are under voluntary control, involving neurons from the brain that descend to the sacral spinal cord. Temperature regulation is coordinated by neural centers in the hypothalamus. In patients suffering spinal-cord injury, however, this supraspinal control can be lost, and reflex responses may be released from higher control.

Similarly, the baroreceptor reflex mechanisms within the brainstem are normally subordinated to activity within higher centers, including areas of the hypothalamus and cerebral cortex. Thus, for example, emotional states, such as anger, can cause a rise of blood pressure that is not corrected by the baroreceptor reflex.

Chemical Transmitters

As discussed in chapter 4 (see figure 4–17), synaptic transmission within the autonomic nervous system involves two types of chemical transmitters: (1) acetylcholine, secreted by cholinergic fibers, and (2) norepinephrine, also known as noradrenaline, secreted by adrenergic fibers. In addition, epinephrine, or adrenaline, is secreted by the adrenal medulla.

Chemical Transmission in the Sympathetic Nervous System

The sympathetic preganglionic fibers secrete acetylcholine at their synapses with postganglionic neurons. The postganglionic fibers in general secrete norepinephrine at their synapses with effectors and are known as sympathetic adrenergic fibers (the sympathetic cholinergic fibers that innervate sweat glands are among the few exceptions to this rule). The sympathetic preganglionic fibers to the adrenal medulla also secrete acetylcholine at their synapses with secretory cells in the adrenal medulla. These cells then secrete a mixture of mostly epinephrine with some norepinephrine into the bloodstream, which carries it to effector organs throughout the body.

Chemical Transmission in the Parasympathetic Nervous System

As in the sympathetic nervous system, the parasympathetic preganglionic fibers secrete acetylcholine. Unlike the sympathetic system, the parasympathetic postganglionic fibers also secrete acetylcholine at their synapses with effectors, and are known as parasympathetic cholinergic fibers.

Receptor Sites

Chemical transmitters react with receptor sites in cell membranes of neurons and of effectors (smooth muscle, cardiac muscle, and exocrine glands). A change in membrane potential and in neural or effector activity follows. The receptor sites are of great practical importance in that numerous drugs and poisons act upon them, and such pharmacological substances are often classified in terms of the appropriate receptor site.

Acetylcholine acts upon two different types of receptor sites. Nicotinic receptors are found in the postganglionic cells within all autonomic ganglia. They are also the receptor type in skeletal muscle membrane. The term *nicotinic* is used because nicotine activates all of these receptors. Muscarinic receptors are found in the effector cells activated by the parasympathetic nervous system and also by the few sympathetic cholinergic fibers. The term *muscarinic* is used because muscarine activates all of these receptors.

Norepinephrine and epinephrine also act upon two different types of receptor sites, both within effector organs activated by sympathetic postganglionic nerve fibers. Alpha receptors cause constriction of blood vessels, relaxation of the intestinal wall, and other effects when activated. They can be stimulated by certain drugs, such as phenylephrine. Beta receptors cause dilation of blood vessels, constriction of the intestinal wall, and other effects when activated. They can be stimulated by certain drugs, such as isoproterenol.

Responses of Effectors

The autonomic nervous system affects many physiological functions. A few important examples are discussed below.

Cardiovascular System

The heart is constantly affected by both sympathetic and parasympathetic neural activity. Sympathetic stimulation causes an increased heart rate and an increased force of contraction, which together lead to an increase in the amount of blood pumped by the heart per minute or an increased cardiac output. Parasympathetic stimulation, on the other hand, causes a decreased heart rate and (in part) a decreased contractile force, leading to a decreased cardiac output.

Sympathetic stimulation causes constriction of the blood vessels that supply the abdominal viscera and the skin. Thus, during the excitement before an athletic event, activation of the sympathetic nervous system can cause not only an increased cardiac output but also can cause an increased fraction of that output to be diverted away from the gut and the skin, to be available to supply the muscles as they prepare for extreme exertion. A dilation of blood vessels that supply skeletal muscle (via stimulation of adrenergic beta receptors), preparatory to exercise, may enhance the diversion of blood flow to muscle.

Gastrointestinal Tract

Parasympathetic stimulation causes increased glandular secretion, peristaltic activity, and sphincter relaxation in the gut; thus, the digestive processes are generally enhanced. Sympathetic activity has opposite effects on peristaltic and sphincter activity, which together with its effect on gastrointestinal blood supply can reduce or shut down the digestive processes.

The Eye

As described in chapter 7, parasympathetic stimulation constricts the pupil while sympathetic stimulation dilates it. Parasympathetic stimulation also constricts the ciliary muscle, leading to accommodation for near vision. These parasympathetic effects are functions of cranial nerve III (oculomotor).

Other Functions

Numerous other functions, described in human and mammalian physiology, are affected by autonomic activity. For example, sexual function is influenced by a combination of parasympathetic and sympathetic activity. Sympathetic stimulation can dilate the bronchi in the lungs, and drugs with a similar effect are often given to relieve asthma attacks. Sympathetic stimulation increases the release of glucose from the liver and the breakdown of muscle glycogen to form glucose and thus tends to increase blood glucose concentration; it also can promote the breakdown of stored fat, releasing free fatty acids into the blood.

Autonomic Denervation

Effector organs depend on their autonomic innervation in a complex way. For example, after the cutting (denervation) of the sympathetic postganglionic nerves to a group of blood vessels, the normal, tonic constriction of the blood vessels will be lost, and they will dilate. After some time, however, a certain degree of constriction will return, because of the activity of the vascular smooth muscle itself. At that time, the effects of sympathetic stimulation produced by injecting norepinephrine or a drug with similar effects will be greatly enhanced, producing more constriction than under normal conditions. This effect is called *denervation supersensitivity* and is attributed to an increase in the number of receptor sites after denervation.

Review Exercises

1. Outline the functional anatomy of the sympathetic and parasympathetic nervous systems.
2. Give two examples of autonomic reflexes. Describe the influence of sensory input and of supraspinal control.
3. Prepare an outline showing the organization of chemical transmission in the autonomic nervous system, including the sites of acetylcholine and norepinephrine secretion, and of nicotinic, muscarinic, alpha, and beta receptors.
4. Briefly describe the effects of sympathetic and parasympathetic stimulation on the cardiovascular system, the gastrointestinal tract, the eye, and the metabolism of glucose.
5. Define denervation and denervation supersensitivity.

11. Cerebral Physiology and Higher Functions

How cerebral activity relates to wakefulness and sleep, memory, language, and other complex processes

In the cockpit of a commercial airliner leaving Baltimore for London, a pilot communicates in a brisk dialogue with the control tower to gain permission to take off. He must be prepared for up to 10 hours of alert wakefulness while paying attention to the information gained through the cockpit window, from the bank of over 100 instruments facing him, and through radio communications. Crucial decisions based on this information, such as whether to take off once maximum ground speed is attained, must often be made within milliseconds. In order to decide properly, important information must be remembered from periods of seconds to hours and evaluated on the basis of criteria learned over thousands of hours of flying time. Wakefulness and alertness, language, and memory are some of the complex physiological and psychological processes that govern how a pilot interprets sensory input and produces the motor output that guides an airplane.

These higher processes are associated particularly with cerebral activity. In some cases, such cerebral activity is reflected in electrical changes, such as the electroencephalogram; in others, it is inferred from the deficits following lesions and functional disorders.

The Electroencephalogram in Wakefulness, Sleep, and Epilepsy

The electroencephalogram, or EEG, is a recording of the electrical activity of the cerebral cortex, normally associated with levels of wakefulness and sleep. Abnormal EEG patterns occur in epilepsy, some other brain disorders, and some types of metabolic dysfunction.

Method

An EEG recording electrode is usually a gold or silver disk or cup about one centimeter in diameter with a wire attached. One electrode, the active electrode, is pasted onto the scalp at some predetermined site over a particular area of cerebral cortex, and another, the reference electrode, is placed on a relatively inactive site, such as an earlobe. The voltage difference between the two electrodes is them amplified and displayed on a chart recorder as a graph against time (figure 11–1). The resulting fluctuations range from 10 to 100 microvolts (millionths of a volt, or 10^{-6} volt) before amplification, and from 0.5 to 50 Hz in frequency. They change with electrode site, age, sensory input and levels of sleep and wakefulness.

In neurological diagnosis, electrodes are usually placed on a number of standardized recording sites. In the widely used 10–20 system of electrode placement, O, P, T, and F stand for occipital, parietal, temporal, and frontal lobe sites, respectively; C stands for the central sulcus region; odd number subscripts stand for left; even for right; and z for midsaggital locations (figure 11–2). Thus, O_1 is a left occipital site and O_2 is the corresponding right occipital site. P_3 and P_4 are corresponding sites over left and right parietal lobes, and so on.

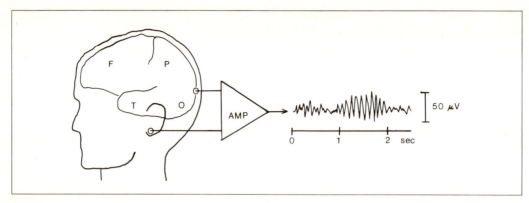

*Figure 11–1. Recording the EEG on a single channel. **AMP** = amplifier, **O** = occipital, **P** = parietal, **T** = temporal, and **F** = frontal.*

The voltage difference between each electrode and the reference electrode is displayed on one of a number of parallel graphs. Alternatively, the voltage difference between two active electrodes, such as O_1 and P_3, can be displayed.

Origin

The EEG obtained from a particular active

Figure 11–2. The International 10–20 electrode system. Source: Jasper, H.H. The ten-twenty electrode system of the International Federation. EEG Clin. Neurophysiol. 10:371–375, 1958. Reprinted with permission.

electrode is believed to originate in a large number (estimates go up to one million) of neurons in the cerebral cortex underlying the electrode site. The small amplitude of the scalp EEG compared with potential fluctuations near the neuron is largely due to the distance and barriers (meninges, blood vessels, bone, skin) between the cortex and the electrode.

Major sources of EEG waves are the postsynaptic potentials (EPSPs and IPSPs) of cortical neurons, particularly from the apical dendrites of cortical pyramidal cells (figure 11–3). When these occur simultaneously in a number of neurons, they summate at the recording site and produce large, regular waves in the EEG, called synchronized EEG waves. Otherwise, the EEG tends to be small in amplitude and

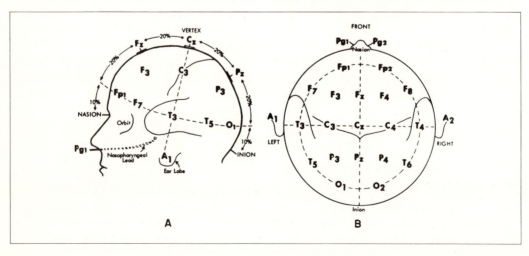

A

B

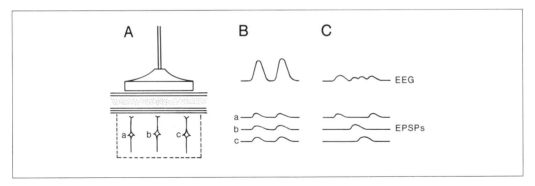

Figure 11–3. Schematic diagram of relationship of EEG with postsynaptic potentials of cortical cells. (A) EEG electrode on scalp can record activity originating in underlying brain area, shown by dashed lines, containing cortical neurons of which a, b, and c, are examples. (B) Synchronized EEG. (C) Desynchronized EEG.

irregular, or desynchronized. The former can be compared to the fluctuations of a sound meter in an auditorium when people are clapping in unison; the latter to the smaller fluctuations when people clap randomly.

Normal EEG Patterns

EEG patterns typically change with levels of wakefulness and sleep in the following manner (figure 11–4):

1. During alertness, with eyes open, waves are low-amplitude, high-frequency, and irregular; they are desynchronized. The frequency range (above 13 Hz) is called *beta*.
2. During relaxation, with eyes closed, waves are of moderately high amplitude, moderate frequency, and regular; they are synchronized. The frequency range (8 to 13 Hz) is called *alpha*. Such waves are not always found in all normal subjects and are most prominent in occipital sites. Opening the eyes makes the waves disappear and shift to the beta pattern, a phenomenon called *alpha blocking*.
3. During most of the night's sleep, the EEG

waves are high-amplitude and low-frequency, including frequency ranges called *theta* (4 to 7 Hz) and *delta* (0.5 to 4). Bursts of waves at about 14 Hz (sleep spindles) may be interspersed during lighter sleep, while slow delta waves predominate during deeper sleep. Because these stages lack rapid eye movements or REMs (see below), they are called *non-REM* or *NREM sleep*.

4. Another stage of sleep is characterized by rapid eye movements (the eyes typically move left and right, back and forth) and is called *REM sleep*. Compared with NREM

Figure 11–4. Normal EEG waveforms. (1) Low-amplitude beta waves during alertness with eyes open. (2) Moderate-amplitude alpha waves, here at a frequency of 10 Hz, during relaxed wakefulness with eyes closed. (3) High-amplitude delta waves during deep, NREM sleep.

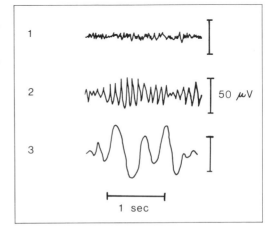

sleep, the EEG is lower in amplitude and higher in frequency (including frequencies in the alpha and beta range). REM sleep has excited a great deal of interest because vivid dreams are often reported by people who are awakened from REM sleep. Several physiological changes, such as loss of postural muscle tone, increased heart and respiratory rate, and activity of the sex organs also occur during this period. In adults, each REM period lasts about 20 minutes and alternates with longer periods of NREM sleep in cycles of about 90 minutes, with three to five of these cycles occurring during the night. Infants spend a larger fraction of their sleep time in the REM stage.

Subcortical EEG Control Centers

What coordinates the millions of neurons throughout the cerebral cortex that give rise to the EEG and associated changes in the level of alertness and sleep? Two centers, the nonspecific thalamic nuclei and the brainstem reticular formation, play major roles. Several other centers are thought to be involved in the onset and regulation of sleep.

Nonspecific Thalamic Nuclei. The appearance of synchronized waves, such as the alpha rhythm, may be related to the activity of neurons in the nonspecific thalamic nuclei. These nuclei include the intralaminar and midline nuclei and project to other thalamic nuclei and to diffuse areas of cerebral cortex, in contrast to specific thalamic nuclei, such as the ventroposterolateral and lateral geniculate, which project to specific, limited areas of cortex (figure 11–5A). The nonspecific projections are via nerve fibers that synapse on superficial cortical layers, particularly onto the apical dendrites of pyramidal cells, while the specific projections synapse more densely on the middle layers. The nonspecific thalamic nuclei and their cortical projections are called the *nonspecific, generalized,* or *diffuse thalamocortical system.*

A proposed model for the origin of the alpha rhythm [1] may be summarized as follows: The alpha rhythm at about 10 Hz is triggered by a similar rhythm of action potentials in nonspecific thalamic neurons. These neurons can be called principal cells since they send axons out of the thalamus to the cerebral cortex [31]. They fire at intervals of about 100 msec (0.1 sec) separated by 100-msec-long IPSPs generated by inhibitory interneurons connected to the principal neurons by a feedback circuit within the thalamus (figure 11–5B). Since the interneurons have numerous branches, they can synchronize the activity of numerous principal cells in the thalamus, which in turn can synchronize the activity of numerous cortical cells. The nerve impulses conducted along principal cell axons to cortical synapses result in depolarizing EPSPs on the cortical surface. The summation of these EPSPs results in a synchronized EEG.

Electrical stimulation of nonspecific thalamic nuclei in animal experiments at about 10 pulses per second has been found to trigger synchronized, depolarizing potentials at the cortical surface, consistent with the role proposed for these nuclei in control of the EEG. Sleep can also result, suggesting a relationship between the synchronized EEG and relaxation or sleep.

Reticular Formation. Desynchronized EEG waves (beta waves) and alert behavior are associated with activity in the brainstem reticular formation, a column of neurons ascending in the central core of the brainstem and containing numerous synapses (figure 11–5A). An early clue to this association was that after epidemics of encephalitis lethargica, or sleeping sickness, a disease often characterized by excessive sleeping and reduced wakefulness, patients were found to have lesions in part of the reticular formation surrounding the cerebral aqueduct in the midbrain (the eye-muscle paralysis often associated with the disease was probably due to invasion of the nearby

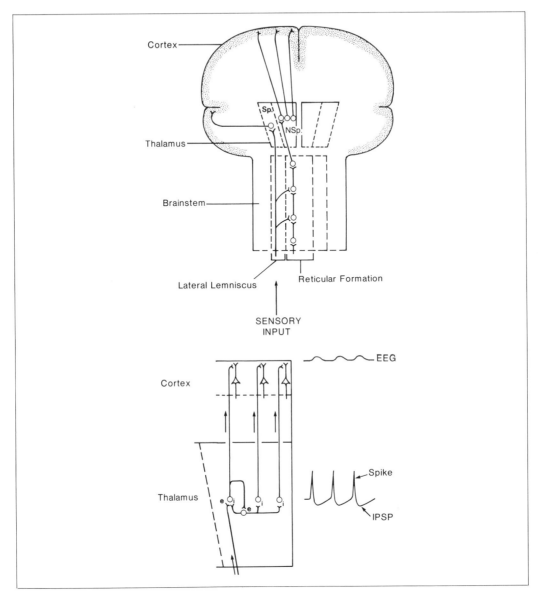

Figure 11–5. (A) *Subcortical areas that coordinate the EEG: the nonspecific thalamic nuclei (**NSp.**), and the reticular formation in the brainstem. Neurons in the nonspecific thalamic nuclei project to diffuse areas of cortex, in contrast to a specific thalamic relay nucleus (**Sp.**), such as the medial geniculate, which projects to a limited sensory area of cortex, such as the primary auditory area. (Synapses of auditory pathway in inferior colliculus not shown.) In the reticular formation, neurons are activated by axon collaterals from the lateral lemniscus* and other sensory pathways. (B) *A model for the origin of the alpha rhythm. Intracellular recordings from principal cells of thalamus (those which project up to cortex) are shown at lower right; **e** = excitation, **i** = inhibition. Sources: G.M. Shepherd. The Synaptic Organization of the Brain (2nd ed.). New York: Oxford University Press, 1979; and P. Andersen and S.A. Andersson. Physiological Basis of the Alpha Rhythm. New York: Appleton-Century-Crofts, 1968.*

oculomotor nuclei). Since then it has been found that damage to the brainstem reticular formation through cerebrovascular accidents, tumors, or increased pressure on the brainstem from other causes can cause loss of consciousness. In comparison, large areas of cerebral cortex can be removed without causing loss of consciousness.

Reticular neurons can be activated by auditory inputs carried up the lateral lemniscus, somatic sensory inputs carried up the medial lemniscus, and by other sensory inputs, all of which converge on reticular neurons (figure 11–5A). Reticular activation experimentally leads to a low-amplitude, high-frequency (desynchronized) EEG, along with alert, wakeful behavior. At least in part, these effects are transmitted by projections from the reticular formation to the nonspecific thalamic nuclei, which then affect cortical activity. Lesions at the upper end of the reticular formation block these effects, resulting in somnolence and a synchronized EEG.

With increasing dosage of general anesthetics [2], both the reticular formation and its projections to the thalamus respond less to sensory stimulation, as shown by neuronal recordings. EEG desynchronization and behavioral alerting are also diminished; these and several other effects of general anesthetics have been attributed to actions on the reticular formation that modify sensory input to the cortex. Direct effects of anesthetics on cortical neurons may also occur.

In addition to its effects on consciousness and the EEG, the brainstem reticular formation has several other functions: it receives projections from the spinothalamic pain pathway (chapter 5), it is a brainstem motor area controlling posture and movement via descending pathways to the spinal cord (chapter 9), and it contains control centers for respiration, heart rate, vasomotor activity and various functions of the autonomic nervous system. That portion of the reticular formation that controls consciousness and EEG activity by means of ascending fibers is known as the *ascending reticular activating system (ARAS).*

Sleep-Inducing Centers. We have already mentioned that low-frequency stimulation of nonspecific thalamic nuclei can result in sleep. In addition, several other areas may be important in inducing sleep. A basal forebrain area, including the anterior part of the hypothalamus and an adjacent area of cerebral cortex, has been found to trigger slow-wave sleep under certain conditions. Animal lesion experiments show a brainstem nucleus called the raphé nucleus, a major source of the presumed neurotransmitter serotonin, to be necessary for slow-wave sleep. The role of serotonin is suggested by experiments in which its synthesis is blocked by drugs, resulting in insomnia. Another brainstem nucleus, the locus ceruleus in the pons, which contains a great deal of norepinephrine, seems to be necessary for REM sleep.

Abnormal EEG Patterns in Epilepsy

The diagnosis of epilepsy is a major clinical application of the EEG. Epilepsy has been defined as "an occasional, an excessive and a disorderly discharge of nervous tissue" [19]. This abnormal electrical discharge occurs in the cerebral cortex or thalamus and (possibly) brainstem reticular formation. The physical or mental manifestations of the abnormal neural activity are known as seizures. Epilepsy may be generalized, involving wide areas of the brain at the same time, or partial, localized to small areas. A text on clinical neurology should be consulted for a more complete treatment of this subject; only some forms of epilepsy are described here, and seizures can occur without epilepsy.

Generalized Epilepsy:

Grand Mal. This is probably the most dramatic form of epilepsy. The patient loses

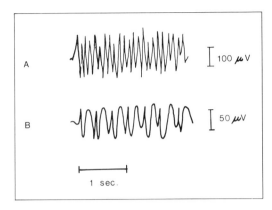

*Figure 11–6. EEG samples during an attack of grand mal (**A**) and petit mal epilepsy (**B**). High-amplitude, repeated spikes are recorded during the tonic phase of a grand mal attack but are frequently obscured by EMG artifact.*

consciousness, falls down, and has convulsive muscle contractions, both continuous (tonic) and jerky (clonic). An EEG recorded during the seizure shows high-amplitude, high-frequency spike potentials or sometimes spikes followed by slow waves, throughout the cortex (figure 11–6A). The probable origin is the brainstem reticular formation and/or the nonspecific thalamic nuclei, which regulate the level of consciousness and influence electrical activity of the entire cortex, as well as (in the case of the reticular formation) influence muscle contraction. These characteristics are summarized in table 11–1, along with those of other forms of epilepsy.

Petit Mal. This involves a brief loss of consciousness, usually lasting less than 60 seconds, with few or no motor signs, and can be so inconspicuous as to go unnoticed. It is often found in children. The very characteristic EEG pattern consists of spike-and-slow wave complexes repeated at about three per second, symmetrical throughout the cortex (figure 11–6B). The probable origin is in the nonspecific thalamic nuclei; artificially induced epileptic

Table 11–1. Some Characteristics of Four Types of Epilepsy

Generalized

Grand mal	Loss of consciousness Tonic and clonic muscle contractions EEG showing repetitive spikes or spike-and-wave complexes throughout cortex Probable origin in brainstem reticular formation and/or nonspecific thalamic nuclei
Petit mal	Brief loss of consciousness Minor motor effects EEG showing spike-and-wave complexes, symmetrical throughout cortex, at about 3 per second Probable origin in nonspecific thalamic nuclei

Partial

Focal motor	Consciousness retained Clonic muscle jerks limited to particular parts of the body EEG showing spikes in motor cortex Probable origin in epileptic focus in motor cortex
Psychomotor	Consciousness altered to variable extent Automatic motor activity often organized into complex acts, and possible mood changes, perceptual distortions, and sensory hallucinations EEG usually shows spikes or slow waves in temporal lobe (but other locations are possible) Probable origin in epileptic focus in temporal lobe

foci in these nuclei have produced a similar syndrome.

Partial (Focal, Local) Epilepsy:

Focal Motor. A typical seizure might start with an involuntary jerk of a finger, while at the same time abnormal spike discharges appear in the hand area of the contralateral motor cortex (the epileptic focus). If the epileptic activity spreads to adjacent areas of the motor cortex, motor activity spreads along corresponding parts of the body (e.g., from finger to hand to arm to the whole side of the body); this phenomenon is known as *Jacksonian March.*

Consciousness may be retained if epileptic activity remains limited to the motor cortex. In some cases, however, epileptic activity spreads from the motor cortex to the nonspecific thalamic nuclei and/or the brainstem reticular formation, leading to a generalized seizure with loss of consciousness as in grand mal epilepsy. The characteristics listed in table 11–1 apply to focal motor epilepsy that does not become generalized.

Psychomotor. The *psycho-* refers to bizarre feelings, mood changes, or hallucinations; a patient may experience feelings of unreality, feelings of anger or fear, or hallucinations, such as events from the past that seem real. The *motor* of psychomotor refers to automatic move-

Figure 11–7. A human somatosensory evoked potential (SEP), recorded with a scalp electrode over the hand projection area of the left postcentral gyrus with reference electrode on left ear, while right median nerve is stimulated. Evoked potential (right) represents average of 120 responses; P25, P45, and P100 are positive peaks at 25, 45, and 100 msec after stimulus onset, respectively. Circled area around scalp electrode is area where P25 is found at maximal amplitude. Source: Adapted from Goff, G.D., Matsumiya, Y., Allison, T., et al. The scale topography of human somatosensory and auditory evoked potentials. EEG Clin. Neurophysiol. 42:57–76, 1977; and Buchsbaum, M.S., Lavine, R.A., Davis, G.C., et al. Effects of lithium on somatosensory evoked potentials and prediction of clinical response in patients with affective illness. In Cooper, T.B., Gershon, S., Kline, N.S., and Schou, M. (Eds.), International Lithium Conference: Controversies and Unresolved Issues. Lawrenceville, N.J.: Excerpta Medica, ICS Series, 1979.

ments, such as buttoning and unbuttoning a jacket. An epileptic focus is generally found in the temporal lobe, so that temporal lobe epilepsy sometimes refers to the same syndrome.

Evoked Potentials

Evoked potentials are voltage changes within the EEG that are evoked or elicited by sudden sensory stimuli, such as electrical stimulation of the finger, a click, or a light flash. To use the first example, an electrical pulse applied to the right index finger so as to give a mild shock sensation sets up nerve impulses in the median nerve that continue up the spinal cord and are relayed from the thalamus to the left somatic sensory cortex (postcentral gyrus). The cortical response recorded by an EEG electrode over that area of the scalp is minute compared with the spontaneous EEG waves. However, if the stimulus is repeated about 100 times and the voltage values at corresponding time intervals after each stimulus are averaged over those repetitions, the cortical sensory response will summate and be enhanced, while the spontaneous EEG waves will tend to cancel each other out. (This technique is known as signal averaging: the sensory response is considered a signal to be extracted from a background of EEG noise).

The resulting waveform is called an *averaged evoked potential* (figure 11–7). It may be compared with evoked potentials recorded directly from the cortex of human or animal subjects. A positive wave that peaks at a latency of 25 to 30 msec after the electrical stimulus is thought to

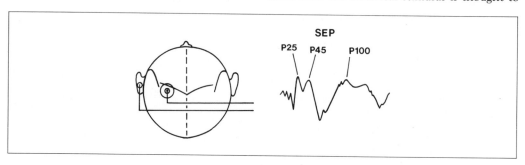

SEP

P25   P45   P100

originate in the synaptic activation of neurons in the middle layers of the postcentral gyrus, where most of the fibers ascending from the thalamic relay nucleus arrive. Later waves (positive peaks at about 50, 100, and 200 msec, and a negative wave at about 140 msec) are thought to be due to less direct pathways to the cortex or to the activation of cortical association areas that surround the primary receiving area in the postcentral gyrus.

In addition to this somatosensory evoked potential (SEP) it is also possible to record an auditory evoked potential (AEP) and a visual evoked potential (VEP). All these responses provide information about the physiological function of human sensory pathways by means of a noninvasive procedure. They have been applied to neurological diagnosis; one example is that reduced conduction velocity in the optic nerve may be associated with increased latency of particular peaks in the visual evoked potential. Other applications are to research in sensory physiology, psychology, and psychiatry. For example, the amplitudes of the late waves are either augmented (increased) or reduced with greater stimulus intensity, and these two response types have shown some correlation with diagnostic categories in psychiatry [5].

Language and Hemispheric Specialization

Language is one of the most uniquely human abilities. It depends on cerebral structures and processes that violate the rule of bilateral symmetry—they are generally concentrated in one cerebral hemisphere rather than the other. The other hemisphere may specialize to some extent in certain higher functions that do not involve language.

Language Abilities and Disorders

Brain mechanisms underlying language were discovered on the basis of brain lesions as-sociated with language disorders. Obviously, oral paralysis interferes with speaking and deafness interferes with hearing words, but there are other types of language disorders, or *aphasias*, that are not due to simple motor or sensory deficits.

The Language-Dominant Hemisphere. Over 90% of the population is right-handed and manipulates pens, forks, and so on under control of the left motor cortex. Almost all of these right-handers also have left hemispheres dominant for language, as indicated by the location of lesions producing language disorders. Of left-handed people, some have left-hemisphere dominance like right-handers, others have cerebral dominance simply reversed so that their right hemisphere is dominant, and a third group have mixed dominance, in which control of language is shared by both sides.

Expressive (Broca's) Aphasia. Paul Broca, in 1861, found that damage to a region of the left frontal lobe gives rise to a disorder in speech production, commonly called expressive aphasia. *Broca's area* (figure 11–8), also called the frontal speech area and including Brodmann's area 44, is adjacent and sends projections to the facial and pharyngeal areas of motor cortex. But with lesions limited to Broca's area, these areas of motor cortex and the corresponding muscles can still function in eating and singing tunes. Speaking may be either absent or limited, labored, and nonfluent, and is often characterized by word-finding difficulties.

Receptive (Wernicke's) Aphasia. In 1874, shortly after Broca's discovery, Carl Wernicke reported that a different lesion in the left hemisphere produced a different language disorder. *Wernicke's area* (figure 11–8), also called the temporal speech area and including part of Brodmann's area 22, adjoins the auditory cortex on top of the temporal lobe and receives

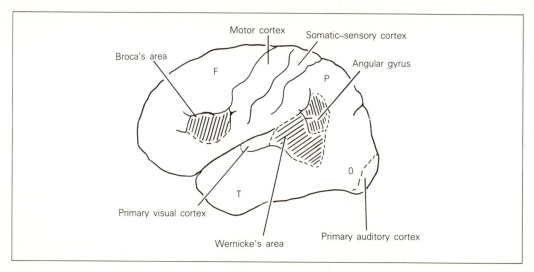

Figure 11–8. Broca's area, Wernicke's area, and the angular gyrus on the left hemisphere. **F, P, T,** *and* **O** *indicate frontal, parietal, temporal, and occipital lobes, respectively. Source: Geschwind, N. Specializations of the human brain. Sci Am. 241:180–201, 1979.*

information about auditory input from it via short nerve fibers. The main language disorder is a difficulty in understanding speech that extends to written language as well. Patients often produce speech fluently but it contains inappropriate words and is relatively incomprehensible; for example: "I came here but the restrictions from the roof falling in on the barn made the army difficult." Because of the patient's difficulty in understanding the meaning of language, he is generally not aware of his own errors.

Alexia. The loss of the ability to comprehend written language due to a brain lesion is called alexia (although inability to read without known brain lesions may be given the same name). When spoken language and calculation ability remain relatively intact, the lesion is generally in the *angular gyrus* in the posterior parietal lobe on the left side (figure 11–8). The angular gyrus is thought to receive visual information from the neighboring visual association cortex and to send information on to Wernicke's area for further processing. Thus, reading a sentence and then repeating it out loud probably involves transmission of neural information from visual cortex, to visual association cortex, to the angular gyrus, to Wernicke's area, to Broca's area, to the oral area of the motor cortex. The pathway from Wernicke's area forward to Broca's area is a fiber bundle known as the *arcuate fasciculus* [9].

Left-Right Hemisphere Specialization. Do lawyers and artists exercise different brain functions in their work? It could be argued that lawyers, using verbal and logical skills involving step-by-step analysis of information, are demonstrating largely left-hemisphere function, while artists, using other abilities, such as perception of visual patterns as a whole, are demonstrating a greater amount of right-hemisphere function. This sort of speculation, although overly simplistic, is one product of studies on left versus right hemisphere specialization. Some methods of study are the following:

1. Patients with unilateral brain lesions have been compared with normal subjects and those with similar lesions on the opposite

side. Lesions in the language areas are examples already discussed.

2. Several patients have had the corpus callosum cut surgically for therapeutic reasons. Subsequent study of the abilities of these "split-brain" patients has led to conclusions about the capacities of left and right hemispheres when not directly connected to each other.

3. In normal as well as brain-damaged subjects, visual input can be limited to the left or right half-field, sending input initially to the opposite hemisphere. Similarly, auditory input can be provided to the left or right ear, sending input primarily to the opposite hemisphere because of the somewhat greater effectiveness of crossed as opposed to uncrossed fibers in the auditory pathway.

4. Cortical electrical activity, both EEG and evoked responses, can be recorded from left and right scalp electrodes. A new method, the positron-emission tomography (PET) scan, provides images of metabolically active areas within the functioning human brain.

The split-brain studies have attracted particular interest (figure 11–9). They utilize the knowledge that the left hemisphere receives input from the right visual half-fields and controls the right arm, while the right hemisphere receives input from the left visual half-fields and controls the left arm. In right-handed patients, when visual input is restricted to the right visual fields and thus the left hemisphere, objects can be named and words can be read aloud or written down, confirming the language-dominance of the left hemisphere. When visual input is switched to the left visual fields and thus the right hemisphere, objects cannot be named and words cannot be read aloud or written down. However, the right hemisphere shows recognition of objects and understanding of spoken and written words to a limited extent as long as linguistic responses are not demanded; for example, objects corres-

ponding to the words can be picked out with the left hand. An explanation is that left-hemisphere input can be processed by centers for language comprehension and production in the same hemisphere; right-hemisphere input can be processed to a limited extent for comprehension in the same hemisphere but cannot cross over to the more capable centers for comprehension and expression in the left hemisphere. Similar tests have shown the right hemisphere to be superior to the left in some nonlinguistic tasks, such as copying a drawing.

Right-Hemisphere Compared with Left-Hemisphere Function. The different methods support the concept that one hemisphere, generally the left, is dominant for language, although they also show a limited amount of language comprehension in the right hemisphere. Arithmetic ability seems to be represented on the same side as language. The results also suggest (although there are some differences of opinion) that the right hemisphere plays a major role in the perception, recognition, and interpretation of visual patterns or objects in space, the person's own body, music, and the meaning of situations.

Lesions in the association cortex of the right posterior parietal lobe (as described in chapter 5) may result in defective perception and neglect of visual objects and of the patient's own body, particularly on the left side. Several neuropsychological tests have shown that problems in visual recognition are more severe following right-sided than left-sided lesions. In normal subjects also, recognition of visual patterns is often superior when they are projected to the right hemisphere rather than to the left hemisphere (from the opposite visual half-fields).

Recent neurophysiological experiments [30] may be related to the right-parietal or neglect syndrome. Single neurons in posterior parietal cortex of monkeys were found to discharge only when a visual stimulus appeared in part of the

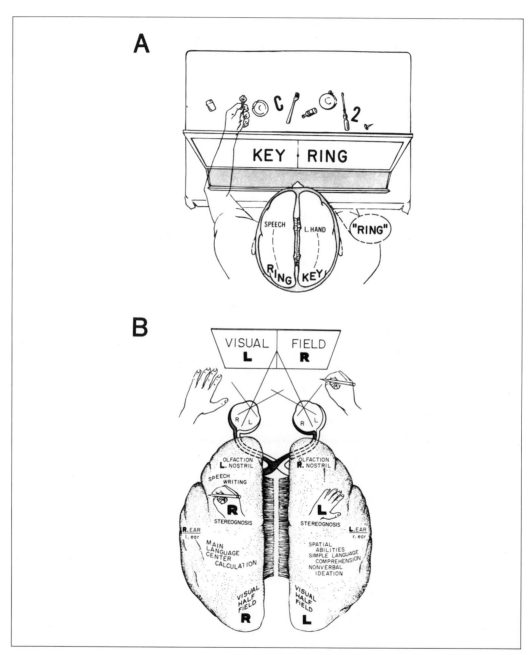

Figure 11–9. (A) *In experiment with split-brain patient, names of objects flashed in left half-field activate right visual cortex and can be read, understood, and picked up by left hand, but cannot be spoken.* (B) *Schema showing visual input, auditory input, hand control, and higher specializations of left and right hemisphere in right-handed subject. Source: Sperry, R. Lateral specialization in the surgically separated hemispheres. In Schmitt, F.O., and Worden, F.G. (Eds.). The Neurosciences: Third Study Program. Cambridge, Mass.: MIT Press, 1974. Reprinted with permission.*

contralateral visual field *and* when the monkey attended to the stimulus, a combination often followed by eye movement to fixate the stimulus. Such neurons may be partly responsible for visual perception associated with attention, and their loss may be associated with visual neglect.

Certain aspects of musical perception have also been related to right-hemisphere function. Patients with right temporal lobe lesions were inferior to patients with left temporal lobe lesions in perception of melody and other nonlinguistic auditory stimuli. In addition, experiments in which separate stimuli entered left and right ears simultaneously (dichotic listening) showed a left-ear superiority for recognizing pitch and chords, implying a right-hemisphere superiority. Similar results have been observed during a test sometimes employed prior to neurosurgery: when the right hemisphere is anesthetized, the patient may be able to speak adequately but unable to hum a tune.

Finally, a right hemisphere role in recognition of the meaning of situations is suggested by several lines of evidence. Patients with right-hemisphere lesions have shown greater difficulty interpreting cartoon sequences and complex pictures than patients with left-hemisphere lesions, although they can identify individual parts of the pictures. They may also deny or minimize their own illness, and display little emotional reaction in general, a tendency called *flat affect*.

Localization and Interrelation of Cerebral Activity. Although a number of higher functions have been localized to particular areas of the cerebrum, the interrelation of brain areas and activities makes the picture more complex [13]. Recovery of language functions after left-hemisphere lesions may reflect compensatory activity of a different part of the brain, for example, the right hemisphere. The effects of apparently similar brain lesions may vary in different patients and under different environmental conditions. Two simultaneous behaviors, such as speaking and tapping, or tapping different rhythms with each hand, may or may not interfere with each other, possibly as a result of the spread of activity between the cortical areas involved. The thalamus and other subcortical regions also play a role in cerebral systems that govern complex behavior. Such examples suggest that cerebral activity in language and other higher processes is complex and our understanding of it incomplete.

Memory

Suppose you have to look up a seven-digit telephone number. If there are no distractions, you may retain the image of the number for an instant after you read it and remember it for a few more seconds while dialing it or repeating it to the operator. If you had to, you could probably memorize the number for a longer period of time, by repeating it over and over again or possibly associating the number sequence with some idea. At any stage, you may well forget the number, especially if you try to memorize several others.

These everyday experiences of memory can be associated with memory processes postulated on the basis of objective psychological experiments on human memory. As diagrammed in figure 11–10 [3,22], visual or auditory input can be thought of as first entering a *sensory store*, where it is stored with little change for up to one second—corresponding in our example to the image of the telephone number. Without attention this information is lost, but with attention it can be transferred to *short-term memory* (STM). Without either effort or distractions, information can remain in STM for 15 to 30 seconds, corresponding in our example to the time it takes to dial or repeat the number. Finally, by repeating (rehearsing) the information several times, or associating it with

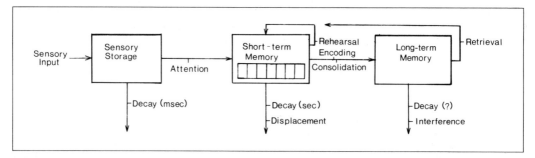

Figure 11–10. A multistage model of human memory. Downward arrows show loss of information. Adapted from Atkinson and Shifrin [3] and Loftus and Loftus [22].

some idea, it can be stored in *long-term memory* (*LTM*) for a period of minutes to years—like your own telephone number and a few others that are important to you.

The stages of sensory store, STM, and LTM were based on experiments on human memory and do not have known anatomical or physiological correlates, although it is possible to make some educated guesses.

Sensory Store

A slide displayed for only 50 msec contains rows of letters, for example,

T   D   R
L   O   N
K   S   T

Shortly after the slide goes off, a tone sounds, signalling the viewer to repeat the letters in one of the rows, corresponding to the pitch of the tone—high, medium, or low. If the delay between the slide and the tone is no more than one second, all the letters can generally be repeated. The explanation has been that an image of the visual stimulus is maintained, probably somewhere in the visual system, for this brief period. A similar sound image is thought to be left by an auditory stimulus. The system postulated for this process has been

called sensory store (sensory storage, iconic store for vision, echoic store for audition).

Short-Term Memory

Although most of the information deluging our sensory store from various sensory inputs is lost, focusing attention on it can transfer some of it to STM. STM can be tested by giving someone a list of digits and asking him to repeat back as many as possible immediately afterward. This procedure, known as a digit-span test, is part of mental-status examinations (given when some impairment of consciousness or thought process is suspected), and of IQ tests, such as the Wechsler Adult Intelligence Scale. Adults can typically repeat back about seven digits correctly (corresponding to the number of digits in a telephone number), with a range of five to nine, or 7 ± 2. The same limit to the capacity of STM applies to letters, words, or other items of information. STM has been compared to a section of computer memory, with five to nine locations containing slots to be filled with items of information.

Although rehearsing (repeating) a short list of items can hold it in STM indefinitely, if rehearsal is prevented (for example, by giving some other mental task immediately after the list is presented) the list will be totally forgotten in 20 to 30 seconds. Thus STM seems to have a built-in limit on both capacity (7 ± 2 items) and duration (20 to 30 seconds).

These limits govern the way information is lost (forgotten) from STM. Because of the limited capacity, new items can enter and displace the old. Because of the limited duration, information can be lost by simple decay if it is not rehearsed or processed further.

Long-Term Memory

How is information transferred from STM to LTM, or *consolidated?* Two thought processes known to be effective are rehearsal and organization. As you might expect, the chance of remembering a single word tends to go up with the number of times the word is rehearsed [22]. For more-complex information, such as a scientific article, memory is aided by organizing the material, with or without rehearsal.

Methods of organizing material may be divided into two categories. The first is by using some preexisting framework. An ancient example, cited by the Roman writer Cicero (in *De oratore*), was the "method of loci." A public speaker would imagine a familiar place, such as his house, and mentally place the parts of his speech in different places—the introduction at the front door, the first anecdote in the living room, and so on. When he gave the speech he would then mentally walk through the house, finding (retrieving) each part of his speech in the corresponding part of the house. Although a preexisting framework is not always available, most people recognize that organizing information themselves, whether by outline, diagram, images, or developing overall ideas about the subject, helps to remember it. Psychology experiments have often used simple lists of unrelated words to study these effects. When asked to memorize ten-word lists, those students who made up a story around the words (for example, "on the *chair* in the kitchen was a *cup,* from which a *canary* drank *water*" used to remember the list *chair, cup, canary, water*) retained most of the words on the list, while others remembered few of the words. Similarly, mental images of parts of such a story helped to remember the words.

Unlike STM, the capacity of LTM is relatively unlimited, or at least unknown, and its duration can last from minutes to a lifetime. A number of experiments suggest that the overall meaning of a written passage rather than the literal words tend to be stored in LTM. Forgetting from LTM is thought to be largely due to interference; for example, memory of one list of words is interfered with by other lists presented before or after the lists in question.

Disorders of Memory

Amnesia is the general term for loss of memory. When amnesia is associated with some clinical condition or episode, it may be retrograde, in which memory is lost for events preceding the episode by a period ranging from minutes to years; anterograde, in which memory is lost for events following the episode; or global, which includes memory loss for events both preceding and following the episode.

Disorders of Memory Associated with Specific Brain Lesions. The clearest association of memory impairment with specific brain lesions has probably been found in patients with lesions of the *hippocampus* and several other structures to which it is anatomically connected [13,23]. The hippocampus is a part of the limbic system within the temporal lobe and adjacent to the lateral ventricle. Several patients sustaining lesions of the hippocampus, often as part of surgical treatment for temporal-lobe epilepsy, have been left with a specific impairment of memory function. Their level of consciousness, concentration, and attention did not seem impaired, their memory for long-past events was good, and they could perform the digit-span test well, even retaining items in memory for several minutes by constant rehearsal. But new information could not be remembered for any longer period if rehearsal was blocked by distractions. In terms of the proposed model, their STM appeared to function adequately, their LTM was largely intact, but consolidation or transfer of information from STM to LTM seemed to be blocked.[1] The result was an inability to retain memory for experiences occurring after the surgery (anterograde amnesia), and sometimes for a period of months to a few years before (a variable amount of retrograde amnesia). These lesions were bilateral (on both sides).

Unilateral lesions of the temporal lobe, including the hippocampus, have been reported to be associated with memory loss for certain information only. These defects correspond to the different functions of the two cerebral hemispheres, in that left-temporal lesions impair verbal memory, while right-temporal lesions impair nonverbal memory, such as recognition of visual designs seen earlier. Although many of these patients had entire temporal lobes removed, the effects of more-limited lesions suggest the critical role of hippocampal lesions in causing the defect.

In patients with Korsakoff's syndrome, a memory disorder accompanies a wide range of conditions including alcoholism, nutritional deficiency, poisoning, and cerebrovascular lesions. Patients have an impaired ability to consolidate new memories (anterograde amnesia) as in hippocampal damage, together with a variable loss of earlier memories (retrograde amnesia), although very early memories tend to be preserved. Lesions have been found in certain thalamic nuclei (the dorsomedial or anterior), a portion of the hypothalamus (the mamillary bodies), and in some cases the hippocampus. These structures are part of a neural circuit projecting from hippocampus via the fornix to the mamillary bodies, from there to thalamic nuclei, and then to the cingulate gyrus of the cerebral cortex.

These cases suggest that the hippocampus, along with some connected structures, plays a role in memory consolidation. This process may be impaired even when consciousness, the

sensory store, short-term memory, and at least a large part of long-term memory appear to function adequately.

Disorders of Memory Associated with More Generalized Brain Dysfunctions. A number of memory disorders are not associated with specific brain lesions, at least according to current knowledge. Amnesia can be induced by a blow to the head (trauma), electroshock, epileptic seizures, cerebral anoxia, or anesthesia. This amnesia is often transient and pertains to events during, shortly before, and sometimes shortly after the episode.

A blow to the head may cause concussion, which typically involves a transient loss of consciousness and of memory for events shortly preceding the injury. Observations of several athletes immediately after concussion in a football game showed that they recalled such events when questioned immediately after the injury (indicating STM function) but were unable to retain the information when questioned 3 to 20 minutes later.

Physiological Mechanisms of Memory

In models of memory based on both human and animal research, short-term memory is thought to involve temporary, easily disrupted changes in neural activity or excitability. One possible change is the activation of closed circuits of neurons, in which, for example, neuron A activates neuron B which then reactivates neuron A, and so on. Such *reverberating circuits* are similar to those thought to occur among spinal interneurons during afterdischarge of the flexor withdrawal reflex; they could be maintained for several seconds but also could be disrupted by another stimulus. Another possible short-term change is found in spinal motoneurons involved in the stretch reflex. If they are activated repetitively at high frequency for a few seconds, a procedure called *tetanic stimulation,* their postsynaptic response

---

[1]There is recent evidence, however, that some skills can be learned, and presentation of cues can improve recall of verbal material [24].

to a single stimulus during a period ranging up to several minutes afterward is increased. This effect is called *post-tetanic potentiation*. It has also been found in various parts of the brain, and is a mechanism by which changes in neuron excitability could be retained for a short period.

Short-term memory has been experimentally disrupted by various treatments that disrupt neural activity. These include electrical shock to the brain (electroconvulsive shock), chemically induced depolarization of cortical neurons (spreading depression), and reduction of temperature (hypothermia) and oxygen supply (anoxia).

Consolidation of information from short-term to long-term memory is often attributed to more stable biochemical and structural changes in neurons, including altered synthesis of RNA and protein, and anatomical changes in synapses. In animals such as fish and mice, the drugs actinomycin D, which interferes with RNA synthesis, and puromycin, which interferes with protein synthesis, appear to block a consolidation stage without interfering with short-term or long-term memory itself. Particularly interesting, in view of the role of the hippocampus in human memory, are the changes found in synaptic structure when a pathway within the hippocampus was stimulated repetitively at high frequency [31]. Not only did post-tetanic potentiation occur, but dendritic spines (small knobs on dendrites which contain postsynaptic terminals) increased in size. Such a change could lead to more effective spread of EPSPs into the postsynaptic cell.

One of the simplest examples of modified behavior or learning is the decline of the response to a regularly repeated stimulus, a change called *habituation* (although there are additional criteria for habituation [20]). In an invertebrate (Aplysia), distinguished by its enormous nerve cells, habituation of a simple behavior, the withdrawal of a gill after sensory stimulation, is retained (remembered) from minutes to hours. This habituation is paralleled by a decline in the EPSP in the motoneuron that supplies the gill muscle. The EPSP decline was attributed to a depletion of transmitter from the presynaptic terminal of the sensory neuron [20].

In mammalian nervous systems, habituation has been found for polysynaptic spinal reflexes, such as the flexor withdrawal reflex. Both a decline in presynaptic transmitter release, as in Aplysia, and a buildup in the activity of inhibitory interneurons have been suggested as possible mechanisms. Single neurons within the brain have also been found to habituate to repeated stimuli. Most of the neurons sampled within the hippocampus appear to habituate, as do a number of neurons sampled within the brainstem reticular formation and the cerebral cortex.

Review Exercises

1. Describe the method of recording the EEG.
2. Diagram and describe a model for the origin of the EEG in cortical neurons and in subcortical EEG control centers.
3. Diagram normal EEG patterns and describe their associatin with varying levels of alertness and sleep.
4. Define REM and NREM sleep.
5. Compare generalized with partial epilepsy. Briefly describe grand mal, petit mal, focal motor, and psychomotor forms of epilepsy.
6. Describe averaged evoked potentials and their relationship with cortical responses to sensory stimulation.
7. Define aphasia and language-dominant hemisphere.
8. Compare expressive with receptive aphasia, in terms of both behavior and brain lesions.
9. Define alexia. Describe its relationship to

the angular gyrus and the interaction of that structure with surrounding areas.

10. List and briefly describe the methods of studying left-right hemisphere specializations.

11. List and briefly describe the higher functions for which either a left- or a right-hemisphere specialization has been reported.

12. Speculate about the probable results of the following experiment. With a group of normal subjects, one word was briefly flashed on a screen in the left visual field, while simultaneously another word was flashed in the right, after which the subjects repeated aloud the word they recog-

nized. Which word was repeated more often in a series of such word-pairs?

13. Diagram a model for human memory including sensory store, short-term memory, and long-term memory.

14. Prepare a table comparing short-term memory with long-term memory in terms of capacity, duration, how information enters, and how information is lost.

15. Define anterograde amnesia and retrograde amnesia.

16. Describe the association of amnesia with brain lesions and brain dysfunctions.

17. Give examples of possible physiological mechanisms underlying short-term and long-term memory.

Glossary

Word origins in Latin (L.) or Greek (Gr.) are given only where particularly helpful for understanding or remembering the term.

Acetylcholine (ACh) Chemical transmitter, synthesized from acetic acid + choline, released by vertebrate motoneurons, preganglionic sympathetic and parasympathetic neurons, and probably other neurons in the central nervous system

Action potential Brief change in membrane potential that is conducted along axon or muscle fiber without decay

Adaptation Diminished response of a sensory neuron to a maintained stimulus

Adiadochokinesis (Gr. *diadochos*, succeeding + *kinesis*, motion) Inability to perform movements that are rapid and alternating

Adrenergic Pertaining to neurons that release norepinephrine (noradrenaline) as a synaptic transmitter

Affect Emotion, outwardly expressed

Afferent (L. *ad*, to + *ferre*, carry) Referring to axons that conduct nerve impulses *toward* the central nervous system, or the impulses themselves

Air conduction Transmission of sound waves to the cochlea via the external auditory canal and the middle ear

Alexia (Gr. *lexis*, word) Loss of reading ability

Alpha blocking Disappearance of alpha waves after eyes are opened or other sensory stimulus is presented

Alpha motoneurons Rapidly-conducting myelinated neurons innervating the extrafusal fibers of skeletal muscles

Alpha receptor A type of adrenergic receptor

Alpha waves EEG waves between 8 and 13 Hz occurring when the subject is relaxed with eyes closed

Amacrine cells One of the five types of retinal cell

Amnesia Loss of memory, of varying extent

Amplitude Height of a wave or signal; a measure of its magnitude

Analgesic (Gr. *angēsis*, pain) Pain-relieving

Angular gyrus Part of the parietal lobe association cortex, important in the comprehension of written words

Annulospiral nerve ending Receptor that signals stretch of a muscle; also termed primary stretch receptor

Anterograde amnesia Defective memory for a period following a brain lesion or other event

Aphasia (Gr. *phasis*, speech) A partial or complete loss of the ability to express or comprehend language (including speech, writing, and signs), generally due to brain pathology; partial loss is also called *dysphasia*

Aqueous humor The fluid in the front part of the eye

Archicerebellum (Gr. *archē*, beginning) A primitive part of the cerebellum, including the flocculonodular lobe, that helps maintain equilibrium

Astigmatism (Gr. *stigma*, point) Refractive error caused by defective curvature of the refracting surface (e.g., cornea), the radius of curvature being greater in one plane than in another perpendicular to it

Athetosis (Gr. *athetos*, not fixed) A movement disorder characterized by slow snakelike, writhing, involuntary movements, often of fingers and toes

Atropine Anticholinergic drug acting at parasympathetic effectors

Audiogram A graph showing hearing threshold as a function of frequency

Averaged evoked potentials CNS response to a sensory stimulus obtained by averaging the EEG data points at corresponding time intervals following a number of stimuli, thereby cancelling those potential oscillations with no fixed relation to the stimulus while enhancing those time-locked to the stimulus

Axon Process or extension of a neuron, capable of conducting nerve impulses to other cells; a nerve fiber

Axon hillock Raised cone-shaped area of the cell body at which the axon originates

Babinski's sign Upgoing of the big toe in response to stroking sole of foot, generally after a lesion of descending motor pathways

Ballismus (Gr. *ballismos*, a jumping about) A movement disorder characterized by violent, flinging, involuntary movements, associated with lesions of the subthalamic nucleus

Basilar membrane Membrane extending longitudinally through cochlea, separating the scala media and the scala tympani and upon which the organ of Corti rests; forms part of the cochlear partition

Best frequency Determined from a tuning curve; the frequency at which a particular neuron in the auditory system has the lowest threshold

Beta receptors A type of adrenergic receptor

Beta waves   EEG waves above 13 Hz

Binocular vision   Simultaneous use of both eyes to yield a single image

Bipolar cells   One of the five types of retinal cell

Blind spot   Small area in the visual field that is unresponsive to visual stimuli, approximately 15 degrees lateral to the fixation point, corresponding to the area of the retina at which the optic nerve exits and photoreceptors are absent

Bone conduction   Transmission of sound waves to the cochlea through the bones of the skull rather than through the middle ear, often used to determine the nature of a hearing loss

Bradykinin   A polypeptide released by tissue damage, causing pain by activation of C fibers

Brainstem auditory evoked potential   Averaged response to auditory stimulus, arising from brainstem auditory pathways

Broca's area   Part of the frontal-lobe association cortex, in front of the facial motor area, important in language expression

Brodmann's areas   A division of cerebral cortex into about 50 areas based on cytoarchitecture

Caloric test   Test of vestibular function by placing warm or cold water in the ear, yielding nystagmus in the normal state

Central sulcus (sulcus of Rolando)   A groove between the motor area in the frontal lobe and the somatic sensory area in the parietal lobe

Cerebral aqueduct   Ventricular passage in the brainstem connecting the third and fourth ventricles

Cerebrum   The main portion of the brain, including the cerebral hemispheres, above the brainstem and cerebellum

Chemoreceptors   Receptors that are responsive to particular molecules either in liquid or gaseous form

Cholinergic   Pertaining to neurons that release acetylcholine as a synaptic transmitter

Chorea   (Gr. *choros*, a dance) A movement disorder characterized by irregular, jerky, dancelike, involuntary movements

Clonus   (Gr. *klonos*, turmoil) Involuntary movement consisting of alternating contraction and relaxation of a muscle group

Coactivation   Acting together of alpha and gamma motoneurons

Cochlea   (L. *cochlea*, snail shell) Fluid-filled, spirally coiled portion of the inner ear, located within the temporal bone and containing the auditory sense organ, the organ of Corti

Complex cells   One type of cell in the visual cortex; it is more responsive to the orientation of a stimulus than to the exact position of the stimulus within the receptive field

Complex periodic sound   Sound composed of several frequencies that yield a complicated but recurrent vibratory pattern

Conduction   Movement of action potential along a nerve or muscle fiber

Cones   Photoreceptors responsible for color (photopic) vision and high visual acuity

Consolidation   A process of fixating information by transferring it from short-term to long-term memory

Convergence   Bringing together to a common point; applied to (1) the refraction of light rays; and (2) the synaptic connections of neurons

Cornea   The transparent structure comprising the foremost portion of the eye

Corpus callosum   Large interconnecting band (commissure) of fibers between left and right cerebral hemispheres

Cortical column   Refers to the cellular organization of the cerebral cortex. In the visual area, cells in a column perpendicular to the surface have the same receptive-field axis orientation; also, at right angles to the orientation columns are alternating ocular-dominance columns each predominantly influenced by one eye.

Corticospinal tract   Tract descending from cerebral cortex through medullary pyramids to spinal cord

Crossed extension   Reflex extension of one leg associated with flexor withdrawal of the opposite leg after a painful stimulus

CT scan   Computerized tomography scan in which the head is scanned with a moving x-ray beam and a series of horizontal sections are reconstructed with computer processing; it provides an image of brain structure based on tissue-density differences

Curare   A drug, extracted from a South American vine, that blocks acetylcholine receptor sites and thereby interferes with neuromuscular transmission

Decibel (dB)   Measure of the intensity of a sound

Deep tendon reflex   *See* **tendon jerk reflex**

Delta waves   EEG waves below 4 Hz

Dendrite   (Gr. *dendritēs*, pertaining to a tree) Process or extension of a neuron, containing postsynaptic sites and capable of receiving neural signals from sensory receptors or other neurons

Denervation supersensitivity   Increased sensitivity

to chemical transmitters seen in muscles, neurons, and glands following denervation

Depolarization Change of membrane potential from resting level toward zero or inside positivity

Dichotic Pertaining to both ears

Divergence Spreading apart from a common point; applied to (1) the refraction of light rays and (2) synaptic connections between neurons

Dominant hemisphere The cerebral hemisphere most essential for language function, generally the left hemisphere

Dopamine A chemical transmitter at some synapses, and a precursor of chemical transmitters (epinephrine and norepinephrine) at other synapses

Dorsal column—medial lemniscus system The somatic sensory pathway that includes dorsal columns, medial lemniscus, VPL thalamic nucleus, and somatic sensory cortex, and is concerned with discriminative touch and pressure, vibration and kinesthesis

Dysarthria (Gr. *arthroun*, utter distinctly) Difficulty in articulating words

Dysmetria (Gr. *metron*, measure) Inability to judge distances during voluntary movement

Edinger-Westphal nucleus In midbrain, parasympathetic component of oculomotor nucleus, governing pupillary constrictor and ciliary muscles of the eye

Efferent (L. *ex*, out + *ferre*, to bear) Referring to axons that conduct impulses *away from* the central nervous system, or the impulses themselves

Electroencephalogram (EEG) Electrical activity of the brain, usually as recorded with electrodes on the scalp

Electromyogram (EMG) Electrical activity arising from muscle activity

Endolymph The fluid filling the scala media of the cochlea and the semicircular canals, similar in composition to intracellular fluid

Endorphin A polypeptide found in brain and pituitary with actions similar to the opiate drugs

End-plate The postsynaptic region of skeletal muscle membrane

End-plate potential (EPP) The depolarization recorded across muscle end-plate membrane following a nerve impulse in the presynaptic motoneuron

Enkephalin Peptide composed of five amino acids and found in brain, with actions similar to the opiate drugs

Epileptic focus Discrete area of cortex from which epileptiform EEG activity originates

Equilibrium potential Membrane potential at which there is equivalent movement of ions in both directions across the membrane, leading to an equilibrium; may be predicted from the Nernst equation

Eustachian tube A tube that connects the middle ear with the nasopharynx and permits equalization of air pressures on both sides of the tympanic membrane

Evoked potential Potential change in the CNS evoked (elicited) by a sensory stimulus, generally recorded with a large electrode

Excitatory postsynaptic potential (EPSP) The potential change generated by excitatory synaptic action in the postsynaptic neuron, tending to excite the postsynaptic neuron

Expressive aphasia A language disorder characterized by defective speech expression

Extrafusal fibers (L. *fusus*, spindle) In skeletal muscle, the contractile fibers that are external to the muscle spindles and make up the bulk of the muscle mass

Extrapyramidal Supraspinal motor areas and their descending tracts that are not part of the pyramidal tracts

Facilitation Increased effect of a stimulus on excitatory synaptic transmission, when another excitatory stimulus is presented (1) shortly before or (2) simultaneously

Fast pain Earliest pain sensation, associated with conduction in A-delta fibers

Feedback Pertains to a system in which some of the output is returned to the input (Examples: hearing oneself speak; baroreceptor response to blood pressure changes due to cardiovascular activity)

Fixation point A point at which one stares to immobilize the eye

Flaccidity Weakness of muscles; lack of muscle tone

Flexor withdrawal reflex Limb withdrawal reflex mediated by nociceptors

Focal length Distance from center of lens to point at which light rays from a distant point converge

Focal motor epilepsy (partial motor seizure) A type of partial epilepsy with motor activity in a particular part of the body

Fovea (fovea centralis) (L. *fovea*, a pit or depression) The depression in the center of the macula lutea of the retina; the area of sharpest vision, upon which the fixation point is normally focused

Frequency   Rate of repetition; the number of cycles made by a sine wave within a second (cycles per second or hertz)

G   Conductance; the ability of ions to pass through a solution or across a membrane within an electric field

GABA (gamma-aminobutyric acid)   A probable chemical transmitter in the CNS

Gamma motoneurons   Small, myelinated neurons innervating intrafusal muscle fibers

Gating   A mechanism by which synaptic transmission in ascending tracts associated with pain sensation may be modulated; associated with the Gate Theory of Melzack and Wall

Generalized epilepsy   A form of epilepsy affecting widespread areas of brain, with seizure activity that is bilaterally symmetrical; includes grand mal and petit mal

Generator potential   Graded, localized electrical potential change in a sensory receptor when properly stimulated; also called a receptor potential

Glycine   An amino acid and probable chemical transmitter in the CNS

Golgi tendon organs   Receptors located in muscle tendons and signaling muscle tension

Graded   Of differing magnitudes

Grand mal epilepsy   (French, major sickness) A type of generalized epilepsy with loss of consciousness, convulsive movements, and postictal (after the fit) depression; EEG initially shows trains of spikes, then spikes and slow waves, and, finally, low-amplitude slow activity

Gyrus   A fold or ridge in the cerebral or cerebellar cortex

Habituation   The gradual waning of a response with repeated stimulation, not due to adaptation of receptors or general fatigue

Helicotrema   Apical end of cochlea to which basilar membrane does not extend; area through which scala vestibuli and scala tympani merge

Hemianopia   (Gr. *hemi*, half + *ope*, vision) Blindness in half the visual field; can involve one or both eyes

Hippocampus   A structure consisting of primitive, three-layered cortex, at the medial edge of the temporal lobe and bordering part of the lateral ventricle; part of the limbic system

Horizontal cells   One of the five types of retinal cell

Hyperopia   (Gr. *hyper*, beyond + *ope*, vision) Visual refractive error in which light rays are brought into focus behind the retina; also called farsightedness

Hyperpolarization   Change of membrane potential from resting level toward a more-negative value (inside of cell referred to outside)

Hypertonus   Excessive muscle tone, with increased resistance to being stretched by an external source

Hypotonus   Diminished muscle tone, with decreased resistance to being stretched by an external source

Hz (hertz)   Cycles per second (c/s)

Inhibitory postsynaptic potential (IPSP)   The potential change generated by inhibitory synaptic action in the postsynaptic neuron, tending to inhibit the postsynaptic neuron

Initial segment   Short, unmyelinated segment of axon just distal to the axon hillock; often the site of impulse origination

Integration   Process by which a neuron combines several excitatory and inhibitory synaptic inputs before generating one or more nerve impulses

Intention tremor   Tremor that is made apparent or aggravated by attempting a coordinated, voluntary movement

Internal capsule   A thick band of white matter in the cerebral hemisphere, containing nerve fibers that connect the cerebral cortex with subcortical structures, such as the thalamus and the spinal cord

Internal environment   The physiological conditions within the body

Interneuron (intrinsic neuron)   A neuron whose short processes remain within a local region of the CNS

Intrafusal fibers   (L. *fusus*, spindle) Thin, specialized muscle fibers within a muscle spindle, containing muscle stretch receptors

Inverse muscle stretch reflex   A reflex, mediated by Golgi tendon organs, which causes a release of tension in a muscle when its tension reaches high levels

Iodopsin   Violet-colored visual pigment found in the retinal cones

Iris   Colored, ring-shaped portion of eye, located behind cornea, containing the muscles by which the pupil is constricted and dilated

Ischemia   Deficiency of blood supply to a part of the body due to reduction of flow through a blood vessel

Isoproterenol   Sympathomimetic agent, which activates beta receptors

Jacksonian epilepsy   A partial motor epilepsy originating in the precentral motor cortex and spreading along the motor area with accompanying motor activity in corresponding parts of the body

Kinesthesis (Gr. *kinesis*, movement + *aesthesis*, perception) The sense of muscle movement and position

Latency (latent period) The time interval between a stimulus and a particular neural or muscular response

Lateral fissure (fissure of Sylvius) A deep groove in the cerebral cortex, partly serving as a boundary for the temporal lobe

Lateral spinothalamic tract A spinal tract ascending to the thalamus, containing nerve fibers responding to pain and temperature stimuli

Law of specific nerve energies Principle that every sensory neuron reacts to only one kind of stimulus and gives rise to only one kind of sensation. Under abnormal circumstances such a neuron can be excited by other stimuli, but it still produces the same sensation; also called Müller's law or the law of specific irritability

L-dopa (L-dihydroxyphenylalanine) A metabolic precursor of dopamine, given as a therapeutic agent in Parkinson's disease

Limbic system (L. *limbus*, border) An interconnected system including the cingulate and parahippocampal gyri, the hippocampal formation, and parts of the amygdala, hypothalamus and thalamus, concerned with visceral and emotional responses and with memory

Locus ceruleus (L. *locus*, place + *coeruleus*, dark blue) A small group of pigmented nerve cells alongside the floor of the fourth ventricle

Long-term memory (long-term store) A hypothetical store or repository of information that is retained for a long period, often years

Macula lutea (L. *macula*, a spot + *lutea*, yellow) A spot at the posterior pole of the eye, containing the fovea centralis at its center

Mechanoreceptor Sensory receptor that responds to mechanical stimuli, such as pressure and displacement

Mental-status examination An evaluation of a patient's mental state, often in standardized form, including assessments of orientation, memory, delusions, hallucinations, emotional state, and cooperation

Middle-ear deafness Deafness resulting from interference with the acoustic transmission of sound within the middle ear; also called conductive hearing loss

Miniature end-plate potential (MEPP) The small depolarization recorded across the muscle end-plate membrane due to spontaneous release of a unit (quantum) of transmitter from the presynaptic motoneruon

Mitral cells The olfactory bulb cells upon which the nerve fibers from the olfactory receptors synapse and whose axons form the olfactory tract

Monosynaptic Having only one synapse in the CNS

Motor cortex In the frontal lobe, precentral gyrus or Brodmann's area 4; origin of corticospinal tract

Muscarinic Pertaining to cholinergic receptors that are activated by muscarine

Muscle spindle Structure found in skeletal muscle, containing intrafusal muscle fibers and receptors for the muscle stretch reflex

Muscle stretch reflex Reflexive muscular contraction elicited by passive longitudinal stretching of the muscle

Muscle tone Steady-state muscle contraction when at rest or when passively stretched by an external force, of reflex origin

Myelin sheath A sheath formed of layers of Schwann or glial cell membranes surrounding myelinated axons, with high resistance and low capacitance

Myopia Visual refractive error in which light rays are brought into focus in front of the retina; also called nearsightedness

Myotatic reflex (Gr. *mys*, muscle + *teinein*, to stretch) The muscle stretch reflex

Narcolepsy A clinical syndrome characterized by attacks of uncontrollable sleep in inappropriate situations, often with rapid onset of REM sleep

Near response Reflex ocular adjustments occurring when gaze is shifted from a far to a near object, including accommodation and pupillary constriction

Neocerebellum (Gr. *neos*, new) The newest part of the cerebellum to appear phylogenetically, receiving major input from the cerebral cortex via the pons, and helping to regulate voluntary movements

Neuromuscular junction Synapse between motoneuron and muscle fiber

Nicotinic Pertaining to those cholinergic receptors that are activated by nicotine

Nociceptors (L. *nocere*, to injure + *capere*, to receive) Sensory receptors that are responsive to potentially injurious stimuli

Nodes Nodes of Ranvier; areas of exposed axonal membrane between myelinated sections

Non-REM (NREM) sleep   Stages of sleep without rapid eye movements; slow-wave sleep

Nonspecific thalamic nuclei   A group of thalamic nuclei, including the midline and intralaminar nuclei, that project to widespread areas of cerebral cortex and exercise a control over their rhythmic electrical activity; the nuclei and their cortical projections are often called the generalized or diffuse thalamocortical system

Norepinephrine (NE, noradrenaline)   Chemical transmitter, synthesized from amino-acid precursor, released by postganglionic sympathetic neurons, adrenal medulla and some neurons in the CNS

Nystagmus   Involuntary, repetitive eye movement in a horizontal, vertical, rotatory (or mixed) direction

Olfactory bulb and tract   The bulb, located on the inferior face of the frontal lobe, is the site of synapse for fibers from the olfactory epithelia and the origin of the olfactory nerve (tract)

Olfactory sensory cells   The receptor organs for the sense of smell

Opiate drugs   A group of drugs, including morphine, codeine, and heroin, that relieve pain as well as causing euphoric and other CNS effects

Optic disk   Point of entrance on retina of optic nerve and blood vessels

Organ of Corti   The sense organ of hearing, located on the basilar membrane within the cochlea, and containing sensory hair cells and auditory nerve endings

Ossicles   The three small, interconnecting bones (maleus, incus, and stapes) of the middle ear which form a system of mechanical levers for efficient transmission of sound energy from the tympanum to the oval window

Oval window   Small, membrane-covered opening into the inner ear, nearly closed by the footplate of the stapes, and the point at which sound energy is transmitted into the inner ear

Pacinian corpuscle   Mechanoreceptor, sensitive to light pressure and vibration, widely distributed throughout the body

Paleocerebellum   (Gr. *palaios*, ancient) A primitive part of the cerebellum, receiving input from the spinocerebellar and other sensory tracts, that helps to control posture and locomotion

Parasympathetic nervous system   Portion of the autonomic nervous system, with efferent nerve fibers originating in certain cranial nerves and sacral segments of the spinal cord

Paresthesia   Abnormal sensation, such as tingling

Parkinson's disease (paralysis agitans)   A motor disorder characterized by a slowing of voluntary movement, masklike face, and resting tremor

Partial epilepsy   A form of epilepsy with seizures beginning locally; focal seizures

Perilymph   Fluid filling the scala vestibuli and scala tympani, nearly identical in composition to cerebrospinal fluid

PET scan   Positron emission tomography scan; a computerized radiographic technique that shows differences in metabolic activity in various areas of the brain

Petit mal epilepsy   (French, minor sickness) Generalized nonconvulsive seizures generally characterized by brief loss of consciousness and 3-per-second spike-and-wave EEG activity

Phase locking   The preferential response (firing) of auditory neurons to one particular phase of the stimulating sound wave

Phasic   (Gr. *phasis*, an appearance) Brief or intermittent

Phenothiazine drugs   A group of drugs, including chlorpromazine, given as therapeutic agents in schizophrenia (and other conditions)

Phenylephrine   Sympathomimetic drug that activates alpha receptors

Photopsin   *See* **iodopsin**

Photoreceptors   Sensory receptors that respond to light—the rods and cones of the eye

Pitch   The property of a sound ranging from high to low and determined by the frequency of the sound waves

Place theory   Theory that the response to a given frequency arises from the place of maximal vertical movement of the basilar membrane for that frequency, the orderly arrangement of such places along the basilar membrane, and the associated neural responses

Plantar reflex   A reflex elicited by stroking the outer aspect of the sole with a blunt object, the normal adult response being flexion of toes and foot

Polysynaptic   Having several synapses in the central nervous system

Positive supporting reaction   A reaction to cutaneous stimulation of the palms of the hands or soles of the feet, involving an extensor thrust

Postcentral gyrus   A fold or ridge immediately behind the central sulcus, containing the primary somatic sensory area

Postganglionic neurons   Autonomic neurons whose cell bodies are in the autonomic ganglia and whose axons terminate upon effector organs

Postsynaptic neuron   The neuron that is excited or

inhibited by another neuron across the synaptic cleft

Post-tetanic potentiation An enhanced neuronal response to a standard stimulus for a prolonged period of time following a tetanic (rapidly repeated) stimulus

Precentral gyrus A fold or ridge immediately in front of the central sulcus, containing the motor area

Preganglionic neurons Autonomic neurons with cell bodies in the CNS and whose axons terminate in the autonomic ganglia

Premotor cortex In frontal lobe, Brodmann's area 6, anterior to motor cortex

Presbyopia (Gr. *presbys*, old + *ōps*, eye) Hyperopia secondary to loss of accommodative power and elasticity of lens, often accompanying aging

Presynaptic inhibition Inhibition arising from the depolarization and consequent reduction of transmitter output of presynaptic nerve endings, resulting from the action of axoaxonal synapses

Presynaptic neuron The neuron that excites or inhibits another neuron across the synaptic cleft

Pretectal nucleus A small nucleus in the midbrain, receiving afferents from the retina and involved in the pupillary light reflex

Principal neuron A neuron that sends a long axon from one brain region to another

Psychomotor epilepsy A type of partial epilepsy combining a disordered mental state and complex motor activity, generally associated with an epileptic focus in the temporal lobe

Pupil The opening in the center of the iris through which light is admitted to the retina

Pupillary light reflex Constriction of the iris in response to light stimulating the retina

Pure tone The auditory signal generated by a sine wave

Pyramidal cell A roughly pyramid-shaped cell in the cerebral cortex, whose axon generally carries output signals from that area of cortex

Pyramidal tract *See* **corticospinal tract**

Raphé nucleus A nucleus in the midline of the brainstem reticular formation, containing a high concentration of serotonin and influencing sleep

Rapid eye movement (REM) sleep A stage of sleep characterized by rapid horizontal eye movements, desynchronized EEG, lowered muscle tone, and frequent reports of dreams

Receptive aphasia A language disorder characterized by defective understanding of speech

Receptive field Area on skin or retina upon which the impingement of an appropriate stimulus will cause the response of a neuron in the sensory pathway

Receptor (1) The peripheral terminal of a sensory neuron or a specialized cell such as a rod or cone, that responds to various environmental stimuli and transduces them into electrical signals; (2) A molecule that combines with a specific chemical substance, such as a transmitter

Receptor potential Membrane potential change in a receptor, caused by an appropriate stimulus; generally a depolarization

Reciprocal inhibition Arrangement of neuronal connections such that excitation of one group of neurons produces inhibition of the antagonistic group

Referred pain Pain sensed as coming from one site but really originating in another site, generally an internal organ

Reflex Involuntary response to a stimulus, involving receptors, afferent impulses to the central nervous system, efferent impulses from the central nervous system, and an effector response. Reflexes are predictable and common to normal members of the species at the same developmental stage

Refraction (1) The bending of light as it passes from one medium into another; (2) The process of determining and correcting refractive errors of the eye

Refractive surface Surface at which light is bent

Refractory period (1) *Absolute.* The period of time following a nerve impulse during which another nerve impulse cannot be generated; (2) *Relative.* The period of time, following the absolute refractory period, during which another nerve impulse can be generated only with a stronger-than-normal stimulus

Repolarization Return of membrane potential from a depolarized state, such as the peak of the action potential, toward the resting potential

Resting potential The steady membrane potential found during the resting state, generally −55 mV to − 90 mV (inside of neuron referred to outside)

Resting tremor Tremor that is present at rest but disappears or diminishes with voluntary movement

Reticular formation (L. *reticulum*, a net) A network of neurons and synapses arranged in a column extending throughout the brainstem in its central region, below the ventricle

Retina Innermost layer of eyeball, containing the photoreceptors

Retinal ganglion cells Final retinal cell layer, the axons of which comprise the optic nerve

Retrograde amnesia Defective memory for a period preceding a brain lesion or other event

Reverberating circuit A closed loop of synaptically-connected neurons around which nerve impulses circulate repetitively after an appropriate input

Rhodopsin The purple visual pigment found in the rods

Rods Most-numerous photoreceptors, responsible for scotopic (night) and peripheral vision and containing the visual pigment rhodopsin

Round window Small membrane-covered opening from middle ear to inner ear, below oval window

Saccule Smaller of two saclike structures in the vestibular apparatus; functions similarly to utricle

Saltatory conduction (L. *saltare*, to jump) Conduction along a myelinated axon, in which the action potential jumps from node to node

Semicircular canal Part of the vestibular apparatus responding to angular acceleration. There are three in each labyrinth, situated at approximate right angles to each other

Sensorineural deafness Deafness secondary to dysfunction of the organ of Corti or the auditory nerve pathway

Sensory store A hypothetical store or repository of information that remains for some milliseconds after a sensory stimulus

Serotonin (5-hydroxytryptamine or 5-HT) A probable chemical transmitter

Servomechanism A device that automatically guides a movement by comparing actual to desired performance and correcting errors

Short-term memory (short-term store) A hypothetical store or repository of information that remains, in general, for a period of seconds

Simple cells One type of cell found in the visual cortex; the receptive field of each has a specific position and orientation

Sine wave A curvilinear waveform consisting of continuous equal alterations of polarity at a set rate

Slow pain Later pain sensation, associated with conduction in C fibers

Snellen test A test of visual acuity

Solitary tract Medullary tract composed of the primary visceral afferent fibers of facial (VII), glossopharyngeal (IX), and vagus (X) nerves

Soma Cell body

Somatic sensory Denoting sensations from the limbs and the body, exclusive of the viscera

Somatotopic mapping The orderly representation of the body on the surface of the somatic-sensory cortex or other level of the somatic-sensory system

Spasticity Hypertonicity

Specific thalamic nuclei A group of thalamic nuclei projecting to specific cortical areas (e.g., the lateral geniculate nucleus, projecting to primary visual cortex)

Spike (1) An action potential; (2) A spike-shaped, sharp wave in an EEG

Split-brain patients Patients in whom the corpus callosum and other commissures between the cerebral hemispheres have been split for therapeutic reasons

Striatum Part of the basal ganglia, including the caudate nucleus and putamen

Substantia nigra A midbrain nucleus with motor functions; its dark appearance is due to melanin pigment

Sulcus A groove in the cerebral cortex

Summation Adding together of subthreshold potentials, either spatially or temporally

Superior colliculus Midbrain structure involved in ocular accommodation for viewing near objects and in reflex movements of the head and eyes for following objects in motion

Supraspinal Above the spinal cord, pertaining to the brain

Surround inhibition The effect of stimulation of an annular (ringlike) area of the skin or retina surrounding an excitatory receptive field, leading to inhibition of a neuron in the somatic sensory or visual system, respectively

Sympathetic nervous system Portion of the autonomic nervous system, with efferent nerve fibers originating in thoracic and upper lumbar segments of the spinal cord

Synapse (Gr. *synaptō*, to join) The site at which two neurons approach each other across a microscopic gap and at which one neuron excites or inhibits the other

Synaptic cleft The microscopic space between presynaptic and postsynaptic membranes

Synaptic potential Change in membrane potential in postsynaptic membrane, caused by synaptic transmission

Synaptic vesicles Small sacs in presynaptic nerve terminals, bounded by membrane and containing chemical transmitter

Synchronized EEG waves Relatively high-amplitude and low-frequency EEG waves

Taste buds The sense organs of taste (gustation)

Temporal lobe epilepsy A type of partial epilepsy arising in the temporal lobe and generally associated with psychomotor seizures

Tendon jerk reflex A phasic muscle stretch reflex elicited by tapping a tendon with a reflex hammer, thereby producing a sudden stretch of the muscle

Thermoreceptors Sensory receptors that respond to changes in temperature

Theta waves EEG waves between 4 and 7 Hz

Threshold (1) The smallest depolarization eliciting an action potential; (2) The smallest sensory stimulus that can be detected

Threshold of hearing The softest sound that can be detected 50% of the time, for a specified frequency

Tonic (Gr. *tonikos*, tension, tone) Pertaining to a chronically active state

Tonic neck reflex (TNR) A reflex response in which forcibly turning the head to one side produces extension of the ipsilateral limbs and flexion of the contralateral limbs

Transducer Structure or device that converts one form of energy into another (e.g., auditory hair cells and microphones both convert sound-induced mechanical vibrations into electrical potential changes)

Tubocurarine A neuromuscular blocking agent, the active principle of curare, that competes with acetylcholine for receptor sites in the muscle end-plate

Tuning curve Graph of threshold sound intensity versus frequency for a neuron in the auditory system

Tympanic membrane (tympanum; eardrum) Cone-shaped membrane that closes the end of the external auditory canal (meatus) and to which the first ossicle, the maleus, attaches

Utricle Larger of two saclike structures in the vestibular apparatus, responding to gravitational pull and acceleration

Visual acuity A measure of the ability of the eyes to resolve detail, distance vision being tested with a Snellen chart and near vision with Jaeger cards

Visual angle The angle subtended between two lines drawn from the eye to the extremities of the object in view

Visual field That area which can be seen by a fixed eye

Vitreous humor The fluid behind the lens of the eye, also called the vitreous body

Volley theory Theory that frequency information is coded by the timing of auditory action potentials

Wernicke's area Part of the temporal-lobe association cortex, behind the auditory cortex, important in language comprehension; may also include adjacent parietal area

References

1. Andersen, P., and Andersson, S.A. *Physiological Basis of the Alpha Rhythm*. New York: Appleton-Century-Crofts, 1968.
2. Angel, A. Effects of anaesthetics on nervous pathways. In T.C. Gray, J.F. Nunn, and J.E. Utting (Eds.), *General Anaesthesia*. 4th ed. Vol. 1. London: Butterworth, 1980.
3. Atkinson, R.C., and Shifrin, R.M. Human memory: A proposed system and its control processes. In K.W. Spence and J.T. Spence (Eds.), *The Psychology of Learning and Information: Advances in Research and Theory*. Vol. 2. New York: Academic Press, 1968.
4. Bosma, J. (Ed.). *Fourth Symposium on Oral Sensation and Perception*. Bethesda: U.S. Department of HEW, 1973.
5. Buchsbaum, M.S., Lavine, R.A., Davis, G.C., Goodwin, F.K., Murphy, D.L., and Post, R.M. Effects of lithium on somatosensory evoked potentials and prediction of clinical response in patients with affective illness. In T.B. Cooper, S. Gershon, N.S. Kline, and M. Shou (Eds.), *International Lithium Conference: Controversies and Unresolved Issues*. Lawrenceville, N.J.: Excerpta Medica, ICS Series, 1979.
6. Dowling, J.E. Organization of vertebrate retinas. *Invest. Ophthalmol.* 9:655–680, 1970.
7. Eccles, J.C., and Sherrington, C.S. Numbers and contraction values of individual motor-units examined in some muscles of the limb. *Proc. Roy. Soc. Lond., Series B*. 106:326–356, 1930.
8. Evarts, E. Brain mechanisms of movement. *Sci. Am.* 241:167–179, 1979.
9. Geschwind, N. Specializations of the human brain. *Sci. Am.* 241:180–201, 1979.
10. Goff, G.D., Matsumiya, Y., Allison, T., et al. The scale topography of human somatosensory and auditory evoked potentials. *EEG Clin. Neurophysiol.* 42:57–76, 1977.
11. Goldstein, M.H., and Abeles, M. Single unit activity of the auditory cortex. In W.D. Kiedel and W.D. Neff (Eds.), *Handbook of Sensory Physiology. Vol. V/2. Auditory System: Physiology, Behavioral Studies, Psychoacoustics*. New York: Springer-Verlag, 1975.
12. Gregory, R.L. *Eye and Brain: The Psychology of Seeing* (3rd ed.). New York: World University Library, McGraw-Hill Book Co., 1977.
13. Hecaen, H., and Albert, M.I. *Human Neuropsychology*. New York: John Wiley & Sons, Inc., 1978.
14. Hille, B. Gating in sodium channels of nerve. *Annu. Rev. Physiol.* 38:139–152, 1978.
15. Hodgkin, A.L. *The Conduction of the Nervous Impulse*. Springfield, Ill.: Charles C Thomas, 1964.
16. Hubel, D.H., and Wiesel, T.N. Receptive fields, binocular interaction and functional architecture in the cat's visual cortex. *J. Physiol.* 160:106–154, 1962.
17. Hubel, D.H., and Wiesel, T.N. Ferrier Lecture: Functional architecture of macaque monkey visual cortex. *Proc. Roy. Soc. Lond, Series B.* 198:1–59, 1977.
18. Hubel, D.H., and Wiesel, T.N. Brain mechanisms of vision. *Sci. Am.* 241:150–162, 1979.
19. Hughlings Jackson, J. A study of convulsions. In Taylor, J. (Ed.), *Selected Writings of John Hughlings Jackson. Volume One: On Epilepsy and Epileptiform Convulsions*. New York: Basic Books, Inc., 1958, p. 8.
20. Kandel, E.R. *Cellular Basis of Behavior*. San Francisco: W.H. Freeman & Co., 1976.
21. Kuffler, S.W., and Nicholls, J.G. *From Neuron to Brain: A Cellular Approach to the Function of the Nervous System*. Sunderland, Mass.: Sinauer Associates, Inc., Publishers, 1976.
22. Loftus, G.R., and Loftus, E.F. *Human Memory: The Processing of Information*. Hillsdale, N.J.: Lawrence Erlbaum Associates, Inc., 1976.
23. Milner, B.R. Amnesia following operation on temporal lobes. In C.W.N. Whitty and O.L. Zangwill (Eds.), *Amnesia*. London: Butterworth, 1966.
24. Oakley, D.A. Brain mechanisms of mammalian memory. *Br. Med. Bull.* 37:175–180, 1981.
25. Palay, S.L. Principles of cellular organization in the nervous system. In G.C. Quarton, T. Melnechuck, and F.O. Schmitt (Eds.), *The Neurosciences: A Study Program*. New York: Rockefeller, 1967.
26. Peper, K., and McMahan, U.J. Distribution of acetylcholine receptors in the vicinity of nerve terminals on skeletal muscles of the frog. *Proc. Roy. Soc. Lond., Series B.* 181:431–440, 1972.
27. Pfaffman, C., et al. Neural mechanisms and behavioral aspects of taste. *Annu. Rev. Psychol.* 30:283–325, 1979.
28. Porter, R. The neurophysiology of movement performance. In C.C. Hunt (Ed.), *MTP International Review of Science, Neurophysiology*. Vol. 3. London: Butterworth, 1975.
29. Rall, W. Theoretical significance of dendritic trees for neuronal input-output relations. In R.F.

Reiss, (Ed.), *Neural Theory and Modelling.* Palo Alto: Stanford University Press, 1964.

30. Robinson, D.L., Goldberg, M.E., and Stanton, G.B. Parietal association cortex in the primate: Sensory mechanisms and behavioral modulations. *J. Neurophysiol.* 41:910–931, 1978.
31. Shepherd, G.M. *The Synaptic Organization of the Brain* (2nd ed.). New York: Oxford University Press, 1979.
32. Smorto, M.P., and Basmajian, J.V. *Clinical Electro-Neurography* (2nd ed.). Baltimore: The Williams & Wilkins Co., 1979.
33. Vallbo, A.B. Muscle spindle response at the onset of isometric voluntary contractions in man. Time difference between fusimotor and skeletomotor effects. *J. Physiol.* 218:405–431, 1971.
34. Willis, W.D., and Grossman, R.G. *Medical Neurobiology* (2nd ed.). St. Louis: The C.V. Mosby Co., 1977.

Suggested Reading

General textbooks

Daube, J.R., Sandok, B.A., Reagan, T.J., and Westmoreland, B.F. *Medical Neurosciences.* Boston: Little, Brown & Co., 1978.

Mountcastle, V.B. (Ed.). *Medical Physiology* (13th ed.). Vol. 1. St. Louis: The C.V. Mosby Co., 1974.

Newman, P.P. *Neurophysiology.* New York: SP Medical & Scientific Books, 1980.

Ruch, F.L., and Patton, H.D. *Physiology and Biophysics* (20th ed.). Philadelphia: W.B. Saunders Co., 1976.

Schmidt, R.F. (Ed.). *Fundamentals of Neurophysiology* (2nd ed.). New York: Springer-Verlag, 1978.

Schmidt, R.F. (Ed.). *Fundamentals of Sensory Physiology.* New York: Springer-Verlag, 1978.

Willis, W.D., and Grossman, R.G. *Medical Neurobiology* (2nd ed.). St. Louis: The C.V. Mosby Co., 1977.

Reading for Chapters 1 through 4

Barr, M.L. *The Human Nervous System: An Anatomic Viewpoint* (3rd ed.). Hagerstown, Md.: Harper & Row Publishers, Inc., 1979.

Brodal, A. *Neurological Anatomy.* New York: Oxford University Press, 1969.

Eccles, J.C. *The Understanding of the Brain.* New York: McGraw-Hill Book Co., 1973.

Gilman, A.G., Goodman, L.S., and Gilman, A. (Eds.). *The Pharmacological Basis of Therapeutics* (6th ed.). New York: Macmillan Publishing Co., Inc. 1980.

Kandel, E.R. *Cellular Basis of Behavior.* San Francisco: W.H. Freeman & Co., 1976.

Katz, B. *Nerve, Muscle, and Synapse.* New York: McGraw-Hill Book Co., 1976.

Kuffler, S.W., and Nicholls, J.G. *From Neuron to Brain: A Cellular Approach to the Function of the Nervous System.* Sunderland, Mass.: Sinauer Associates, Inc., Publishers, 1976.

Noback, C.R., and Demarest, R.J. *The Human Nervous System. Basic Principles of Neurobiology* (2nd ed.). New York: McGraw-Hill, 1977.

Shepherd, G.M. *The Synaptic Organization of the Brain* (2nd ed.). New York: Oxford University Press, 1979.

Snell, R. *Clinical Neuroanatomy for Medical Students.* Boston: Little, Brown & Co., 1980.

Reading for Chapters 5 through 8

Bonica, J.J. (Ed.). *International Symposium on Pain.* New York: Raven Press, 1979.

Brindley, G.S. *Physiology of the Retina and the Visual Pathway.* London: Edward Arnold. 1970.

Dallos, P. *The Auditory Periphery.* New York-London: Academic Press, 1973.

Davson, H. *The Physiology of the Eye* (3rd ed.). New York: Academic Press, 1972.

Evans, E.F., and Wilson, J.P. *Psychophysics and Physiology of Hearing.* London: Academic Press, 1977.

Gregory, R.L. *Eye and Brain: The Psychology of Seeing* (3rd ed.). New York: World University Library, McGraw-Hill Book Co., 1977.

Kenshalo, D.R. (Ed.). *The Skin Senses.* Springfield, Ill.: Charles C Thomas, 1968.

Reading for Chapter 9

Granit, R. *The Basis of Motor Control.* London: Academic Press, 1970.

Magoun, H.W., and Rhines, R. *Spasticity: The Stretch Reflex and Extrapyramidal System.* Springfield, Ill.: Charles C Thomas, 1947.

Phillips, C.G., and Porter, R. *Corticospinal Neurones.* New York: Academic Press, 1977.

Reading for Chapter 10

Appenzeller, O. *The Autonomic Nervous System: An Introduction to Basic and Clinical Concepts.* New York: American Elsevier Publishers Inc., 1970.

Gilman, A.G., Goodman, L.S., and Gilman, A. (Eds.). *The Pharmacological Basis of Therapeutics* (6th ed.). New York: Macmillan Publishing Co., Inc., 1980.

Ryall, R.W. *Mechanisms of Drug Action on the Nervous System.* Cambridge: Cambridge University Press, 1979.

Reading for Chapter 11

Hecaen, H., and Albert, M.I. *Human Neuropsychology.* New York: John Wiley & Sons, 1978.

Kooi, K. *Fundamentals of Electroencephalography* (2nd ed.). New York: Harper & Row Publishers Inc., 1978.

Loftus, G.R., and Loftus, E.F. *Human Memory: The Processing of Information.* Hillsdale, N.J.: Lawrence Erlbaum Associates, inc., 1976.

Luria, A.R. *The Working Brain.* London: Penguin, 1974.

Handbooks and Reviews

Brookhart, J.M., and Mountcastle, V.B. (Eds.). *Handbook of Physiology.* Section 1. *The Nervous System.* (*a*) Volume 1: Kandel, E.R. (Ed.), *Cellular Biology of Neurons.* Bethesda, Md.: American Physiological Society, 1977. (*b*)
Volume 2: Brooks, V.B., and Geiger, S.R. (Eds.), *Motor Control.* Bethesda, Md.: American Physiological Society, 1981.

Field, J., Magoun, E.W., and Hall, V.E. (Eds.). *Handbook of Physiology.* Section 1: *Neurophysiology.* 3 vols. Washington, D.C.: American Physiological Society, 1959. (*a*) Kandel, E.R. (Ed.), *Ceullar Biology of Neurons.* Vol. 1, 1977. (*b*) Brooks, V.B. (Ed.), *Motor Control.* Vol. 2, 1981.

Handbook of Physiology. Section 1: *Neurophysiology.* Washington, D.C.: American Physiological Society, 1959.

Handbook of Sensory Physiology. New York: Springer Publishing Co., Inc., 1971.

Gazzaniga, M.S., and Blakemore, C. (Eds.). *Handbook of Psychobiology.* New York: Academic Press, 1975.

Hunt, C.C. (Ed.). *MTP International Review of Science,* Ser. 1: *Physiology.* Vol. 3: *Neurophysiology.* Baltimore: University Park Press, 1975.

Porter, R. (Ed.). *International Review of Physiology.* Vol. 10. *Neurophysiology II,* 1976. Vol. 17. *Neurophysiology III,* 1978. Baltimore: University Park Press, 1976, 1978.

Periodicals containing up-to-date reviews of specific topics include *Annual Review of Physiology, Annual Review of Psychology, Annual Review of Pharmacology, International Review of Neurobiology,* and *Physiological Reviews.*

Index

Accommodation, 91–92
Acetylcholine, 48–50
  in autonomic nervous system, 136–137
  in basal ganglia, 125
  in motoneuron, 38
  receptors, 39
Action potential, 8, 24–29
  all-or-none law of, 25
  conduction of, 27–28
  duration of, 10
  frequency of, 11
  ionic basis of, 25
  in receptors, 33
  recording of, 28–29
  threshold of, 24
  time intervals of, 11
  velocity of, 10–11, 28, 55–57
Adaptation, 33–35
Adrenal medulla, 134–135
Adrenergic neurons, 51
Air conduction, 80
Alexia, 148
Alpha blocking, 141
Alpha-bungarotoxin, 41
Alpha motoneurons, 116–117
Alpha receptors, 137
Alpha waves, 141–142
Amacrine cells, 95
Amnesia, 153
Amygdala, 8
Anticholinesterase, 52–53
Aphasia, 147–148
Arcuate fasciculus, 148
Ascending reticular activating system
  (ARAS), 144
Astigmatism, 90–91
Audiogram, 79–80
Audiometry, 79–80
Audition
  bone conduction, 71
  cochlear mechanics, 72
  hearing sensitivity and loss, 78–82
  middle ear functions, 70–71

Audition (continued)
  neuronal responses, 74–78
  receptors, 73–74
  sound waves, 69–70
Auditory cortex, 77–78
Autonomic nervous system, 13–14, 133–138
  chemical transmitters, 136–137
  denervation, 138
  effector responses, 137–138
  reflexes, 136–137
  supraspinal control of, 136
Axon, 1

Babinski's sign, 121, 123
Basal ganglia, 8, 125–127
Best frequency, 74
Beta receptors, 137
Beta waves, 141
Binocular vision, 103–104
Bipolar cells, 94–95
Blind spot, 87
Bone conduction, 71
Botulinum toxin, 52
Brainstem, 7
  auditory evoked response, 80–82
  motor areas, 123–125
Broca's area, 147–148

Catechol-O-methyl transferase (COMT), 50
Cell body, 1
Central nervous system (CNS), 2, 6–8
Cerebellum, 7, 127–130
Cerebral physiology
  electroencephalogram and, 14, 139–147
  epilepsy, 144–146
  evoked potentials, 146–147
  language, 14, 147–151
  memory, 14–15, 151–155
Cerebrum, 8
Channels, 17–19, 27
Chemical transmitters, 10, 38–39, 48–53,
  125, 136–137
Cholinergic neurons, 50

Clonus, 118
Coactivation, 118, 123
Cochlea
  microphonic, 74
  structure, 72
  traveling waves, 72–73
Color blindness, 104
Color sensitivity, 98–99
Color vision, 104
Compound action potential, 29
Conduction
  of action potential, 27–28
  and middle-ear deafness, 80
  in motor processes, 13
  in myelinated axons, 27
  in sensory processes, 12
  in unmyelinated axons, 27
Conduction velocity
  of alpha motoneurons, 114
  of Group Ia fibers, 114
  in myelinated and unmyelinated nerve fi-
    bers, 28
  in sensory nerve fibers, 56
Cones, 94, 97–99, 104
Consciousness, 14, 144
Convergence, of light rays, 89
Corpus callosum, 8
Corticospinal tract, 121–122
Curare, 40–41, 52

Dark adaptation, 98
Decibels, 79
Delta waves, 141
Demyelination, 28
Dendrites, 1
Denervation supersensitivity, 138
Depolarization, 23, 26, 45
Deuteranopia, 104
Digit span test, 152
Dopamine, 51, 125–126
Dorsal column system
  and CNS activity, 58–59
  and discriminative somatic sensation,
    58–65
  and peripheral nerve activity, 58

Dorsal column system (*continued*)
  receptive fields in, 60
  and somatotopic mapping, 59–60
  and surround inhibition, 60–61
Drug action, 52–53

Edge enhancement, 97
Effectors, 1
Electrical properties, 21–23
Electroencephalogram (EEG), 14
  method, 139
  normal patterns in, 141–142
  origin of, 140–141
  subcortical control centers, 142–144
  10–20 system in, 139–140
Electromyogram (EMG), 113
End-plate potential (EPP), 39
Epilepsy
  and EEG, 144–146
  focal motor, 145–146
  generalized, 144–145
  grand mal, 144–145
  Jacksonian March, 145–146
  partial, 145–146
  petit mal, 145
  psychomotor, 146
  temporal lobe, 146
Evoked potentials, 146–147
Excitatory postsynaptic potential (EPSP),
  42–43, 45
Excitatory synapses, transmission at, 41–42
Extrapyramidal pathways, 124, 129
Eye, as optical system, 87–93

Final common path, 129
Fixation point, 87
Flexor withdrawal reflex, 119
  divergence in, 120
  habituation of, 155
  neuronal mechanism of, 120
  polysynaptic nature of, 120
Focal length, 89–90
Fovea, 87
Frontal lobe, 8

GABA (gamma-aminobutyric acid), 52
Gamma motoneurons, 117
Gating, 65
Gating current, 27
Generator potential, 31
Glycine, 52
Goldman equation, 23
Golgi tendon organ reflex, 118–119
Graded response, 9–10, 32, 34, 43, 44
Gray areas, 7

Habituation, 155
Hair-follicle receptors, 34
Hearing
  loss, 80
  sensitivity, 78–79
  theories of, 74
  theshold of, 78–79
Hemianopia, 99–100
Hemispheric specialization, 148–149
Higher processes, 14
Hippocampus, 153–154
Horizontal cells, 94–95
Hyperopia, 90
Hyperpolarization, 23
Hypothalamus, 7–8

Inhibition, 44–45, 47–48, 61–62, 95–96
Inhibitory interneuron, 119, 142
Inhibitory postsynaptic potential (IPSP), 44
Inhibitory synapses, 44–47
Initial segment, 1
Integration, 12, 13
Inverse stretch reflex, 118–119

Landolt C test, 96
Language, 147–151
  and dominant hemisphere, 147
  and left-right cerebral differences, 14
Latent period, 113–114
Lateral geniculate nucleus, 99–100
Lateral inhibition, 95–96
Light sensitivity, 98–99
Localization, of sound, 75–76
Locus ceruleus, and sleep, 144

Long-term memory (LTM), 15, 151–153, 155

Mechanoreceptors, 31
Meissner's corpuscles, 33–34
Membrane potential, 23
Memory, 14–15, 151–155
  and biochemical changes, 155
  consolidation, 153–155
  disorders, 153–154
  and hippocampus, 153–154
  and Korsakoff's syndrome, 154
  physiological mechanisms of, 154–155
  and structural changes, 155
  and temporal lobe, 154
Microelectrodes, 23
Middle ear
  functions of, 70–71
  deafness, 80
Miniature end-plate potentials (MEPPs), 38–39
Monoamine oxidase (MAO), 51
Motor cortex, 121–123
Motor processes. *See* Posture and movement
Motor unit, 37–38
Muscarinic receptors, 137
Muscle stretch reflex, 113–118
  clinical examination of, 118
  neuronal mechanism of, 115–117
  phasic, 118
  supraspinal control of, 117–118
  tendon jerk, 118
  tonic, 118
Muscle tone, 118
Myasthenia gravis, 53
Myelin sheath, 1
Myopia, 90

Near response, 93
Neglect syndrome, 149–151
Neostigmine, 53
Nernst equilibrium potential, 20–21, 26
Nerve fibers. *See also* Conduction; Conduction velocity
  afferent, 2, 4

Nerve fibers (*continued*)
  alpha motoneuron, 116–117
  classification of, 56–57
  efferent, 2, 4
  gamma motoneuron, 117
  Group Ia, 116
  Group Ib, 119
  Group II, 57–58
  myelinated, 57, 63
  somatic, 2, 4
  unmyelinated, 57, 63
  visceral, 2, 4
Nerve impulse. *See also* Action potential
  and action potential, 24–29
  and membrane potential changes, 23–24
  and nerve membrane, 17–20
  and resting potential, 20–23
Nerve membrane, 17–20
Neuromuscular junction, 1, 37
Neuromuscular transmission, 37–41
Neurons, 1
Nicotinic receptors, 137
Noise, 70
Nonspecific thalamic nuclei, 64, 142
Norepinephrine, 50–51, 136–137
Nystagmus, 83–84

Occipital lobe, 8
Olfactory system, 110–111
Optics. *See* Vision
Orientation column, 102–103
Overshoot, 26

Pacinian corpuscle, 31–33
Pain
  pharmacological relief of, 65–66
  psychophysiological relief of, 66
  referred, 65
  and spinothalamic system, 63–65
Parasympathetic nervous system, 14,
    135–136
Parietal association cortex, 62–63, 149–151
Parietal lobe, 8
Peripheral nervous system (PNS), 2

Phase locking, 75
Place theory of hearing, 73
Plantar reflex, 121
Pores. *See* Channels
Positive supporting reaction, 121
Positron emission tomography, 149
Postsynaptic inhibition, 44, 61
Post-tetanic potentiation, 154–155
Posture and movement, 12–13
  basal ganglia in, 125–127
  brainstem motor areas, 123–125
  cerebellum, 127–129
  flexor withdrawal reflex, 119–120
  inverse stretch reflex, 118–119
  motor cortex in, 121–123
  muscle stretch reflex, 113–118
  overview, 129–130
  spinal reflexes in, 113–121
  supraspinal control of, 121–130
Potassium channels, 20, 23, 26, 27
Potassium ions, 17–21, 22–23, 26
Potentials
  action, 8, 10–11, 24–29, 33
  end-plate, 39
  evoked, 146–147
  generator, 31
  membrane, 8, 23
  miniature end-plate, 38–39
  postsynaptic, 42–44
  receptor, 9, 31–33
  resting, 8, 20–23
Presbyopia, 92
Presynaptic inhibition, 47–48, 61–62
Protanopia, 104
Pupillary constriction and dilation, 92–93
Pyramidal calls, 121
Pyramidal tract. *See* Corticospinal tract

Raphé nucleus, 144
Receptive fields
  of retinal ganglion cells, 95–97
  of visual cortex cells, 100–104
  of somatic-sensory cells, 60–62
Receptor potential, 1, 9, 31–33

Receptor sites, 137
Receptors, 1, 9, 11–12, 31–36. *See also*
    Chemical transmitters; Sensory
    reception
  auditory, 73–74
  mechano-, 31–35
  olfactory, 110–111
  pain, 35, 63
  somatic-sensory, 31–36
  stretch, 34–35, 113–115
  taste, 107
  vestibular, 82–83, 84
  visual, 94, 97–99, 104
Reciprocal inhibition, 115
Red nucleus, 123
Reflex, latent period of, 113–114
Reflexes
  autonomic, 136
  baroreceptor, 136
  flexor withdrawal, 119–120
  Golgi tendon organ, 118–119
  inverse stretch, 118–119
  middle-ear, 71
  muscle stretch, 113–118
  plantar, 121
  pupillary, 92–93
  tonic neck, 121
Refraction, 88–90
Refractive errors, 90–91
Refractive power, 89–90
Refractory period, 26
Repolarization, 26
Resting potential, 8, 20–23
Reticular formation, 123–124, 142–144
Retina, 94–96
Retinal ganglion cells, 94–95
Reverberating circuits, 154
Rhodopsin, 94, 98
Right-hemisphere function, 149–151
Rods, 94, 97–99
Roots, spinal nerve, 6

Safety factor, 28
Saltatory conduction, 28

Semicircular canals, 82–84
Sensorineural deafness, 80
Sensory nerve fibers, 55–57
Sensory processes, 11–12
Sensory reception, 1, 9, 11–12, 31–36. *See
    also* Receptors
  action potential, 33–34
  adaptation, 33–34
  receptor potential, 31–34
  transduction, 12, 31–34
Sensory store, 151–152
Serotonin, 51–52
Short-term memory (STM), 15, 151–153,
    155
Simple cells, 100–103
Sleep
  and locus ceruleus, 144
  non-REM, 141–142
  and raphé nucleus, 144
  REM, 141–142
Smell, 109–111. *See* Olfactory
Snellen Chart, 96–97
Sodium channels, 25–27
Sodium ions, 17
Sodium–potassium pump, 17
Somatic sensation
  discriminative, and dorsal column system,
    58–63
  pain, and spinothalamic system, 63–66
  parietal association cortex, 62–63,
    149–151
  sensory nerve fibers, 55–57, 58, 63
Sound
  localization, 75–76
  waves, 69–70
Spasticity, 118
Specificity, in sensory processes, 12
Spinothalamic system, 63–65
Split-brain patients, 149
Stretch receptor, 34–35, 115–116, 117
Striate cortex, 100
Structure and function
  cellular, 1–2
  of CNS, 6–8

Structure and function (*continued*)
    of higher processes, 14–15
    of motor processes, 12–14
    of neural signals, 8–11
    of PNS, 2–6
    of sensory processes, 11–12
Substantia nigra, 125–126
Summation, 32, 34, 43–44
Supraspinal control, 121–130
Surround inhibition, 95–96
Sympathetic nervous system, 145, 133–135
Synapses. *See* Synaptic transmission
Synaptic delay, 39, 43
Synaptic potentials, 10, 38–40, 42–47
Synaptic transmission, 1, 10
    chemical transmitters in, 48–53
    drug action and, 52–53
    excitatory, 41–44
    inhibitory, postsynaptic, 44–47, 61
    inhibitory, presynaptic, 47–48, 61–62
    neuromuscular, 37–41
    neuron–neuron, 41–48

Taste
    action potentials in, 108
    behavioral and reflex responses, 107–109
    buds, 107
    preferences and aversions, 109
    receptor potential in, 108
    sensation, 107–109
    sensory cells, 107
Temporal lobe, 8
Tetraethylammonium, 27
Tetrodotoxin, 27
Thalamus, 7–8, 59, 63–64
Theta waves, 141

Threshold
    of action potential, 24, 33, 34, 43
    of hearing, 78–79
Tones, as stimuli for hearing, 69
Tonic neck reflex, 121
Touch and pressure, 58
Transduction, 12, 13, 31–34, 73–74
Traveling waves, 72–73
Tricyclic drugs, 53
Tritanopia, 104
Tubocurarine, 41, 52
Tuning curve, 74–75
Two-point tactile discrimination, 58

Vesicles, 1, 37–39, 41, 48, 50–51
Vestibular system, 82–84, 124
Vibratory sense, 58
Vision
    accommodation, 91–92
    acuity, 96–99
    binocular, 103–104
    color, 104
    cortex, 100–103
    edge enhancement, 97
    field of, 87
    neurophysiology of, 93–104
    optics, 87–104
    pathway, 99–100
    pupil size, 92–93
    refraction and refractive errors, 88–91
    retina, 94–99
    rods and cones, 97–99

Wernicke's area, 147–148
White areas, 7